T0290994

Manufacturing Excellence in Spinning Mills

Manufacturing Excellence in Spinning Mills

A. Kanthimathinathan

CRC Press is an imprint of the
Taylor & Francis Group, an **informa** business

First edition published 2022

by CRC Press
6000 Broken Sound Parkway NW, Suite 300, Boca Raton, FL
33487-2742

and by CRC Press
4 Park Square, Milton Park, Abingdon, Oxon, OX14 4RN

CRC Press is an imprint of Taylor & Francis Group, LLC

Library of Congress Cataloging-in-Publication Data
A catalog record has been requested for this book

ISBN: 978-1-032-19100-3 (hbk)
ISBN: 978-1-032-19101-0 (pbk)
ISBN: 978-1-003-25767-7 (ebk)

DOI: 10.1201/9781003257677

Typeset in Palatino
by MPS Limited, Dehradun

Destiny smiles on an individual through benevolent seniors/leaders in the industry. Long years of their insightful and masterly guidance has chiseled the "raw-me" into a "professional-me." I dedicate my work to my icons to convey my gratitude for the elegant and much-enjoyed metamorphosis.

A. Kanthimathinathan

Contents

Foreword

Mr. Ashwin Chandran.
Chairman Managing Director

In this book, the author, who has more than four decades of experience in spinning mills and research institutions has shared his experience in *Quality Assurance, Production Management and Maintenance Management* of modern machinery and laboratory equipment's towards achieving *Manufacturing Excellence.*

Employees play a vital role in helping any organization to achieve its goals and objectives and perform efficiently and optimally. Hence, a separate chapter is devoted to the *Training and Development of employees,* from shop floor employees to senior level managers.

There are also chapters on *Energy Management and Customer Focus,* without which the objectives of the book will not be fulfilled. The author has clearly outlined what should be done to become a ***"Smart Spinner"*** as a pathway towards "Industry 4.0" and smart manufacturing. Spinning mills can derive a lot of inputs from this book that will enable them to move forward towards excellence in manufacturing.

The author has derived from the wealth of his industry and research experiences to put together this book which will give entrepreneurs and spinning mill promoters the tools required to achieve excellence in manufacturing and ensure that their businesses will remain sustainable and competitive. This book will also serve as reference document for all Technicians and Technologists who work in spinning mills. Moreover, academic institutions will gain by adding this book to their curriculum to emphasize the values of Quality Assurance, Maintenance Management, Production Management, Customer Focus, Training and Development, Energy Management and "Industry 4.0" in spinning mills.

I commend the author for bringing out this valuable work which will benefit the spinning industry and wish him all the best in this and future endeavours.

Ashwin Chandran

Precot Limited (Formerly Precot Meridian Limited)
Regd Office, D Block, 4th Floor, Hanudev Info Park, Nava India Road,
Udaiyampalayam, Coimbatore - 641 028
Tel: 0091 422 4321100 | Fax: 0091 422 4321200 | Email: co@precot.com
CIN: L17111TZ1962PLC001183 | Website: www.precot.com

Preface

Achieving manufacturing excellence in any organization offers a wide range of benefits to it over anyone running the business "as usual." In a deeper sense, excellence in manufacturing helps the organization to "delight their customers, cut costs by controlling the draining of income, improve throughput, effective utilization of all the resources and also to safeguard the interests of the stakeholders."

The following strategic steps are suggested to achieve manufacturing excellence:

1. Business-process reengineering: Question the status quo and improvise all activities in the organization.
2. Create measures of performance metrics to evaluate the manufacturing system.
3. Identify and eliminate production bottlenecks.
4. Audit to identify where the resources are wasted and establish systems to eliminate permanently, wherever possible.
5. Organize everything so as to make it transparent and accessible/create standard operating procedures.
6. Evaluate the skills and competence levels of every employee and set up a full-fledged "training system."
7. Establish proactive/predictive management of equipments/machines.
8. Create vendor and customer assessment systems/create customer satisfaction index.
9. Work toward smart manufacturing by aligning with the evolution of Industry 4.0.

Spinning mills follow different kinds of systems as discussed earlier in totality or partially; but, those who are outperforming their peers certainly follow all these systems and continue to improve further in pursuit of excellence.

This executive handbook on *Manufacturing Excellence in Spinning Mills* is an effort by the author to bring out the critical and important elements of manufacturing operations quality, maintenance, production, training, energy management, and customer focus so that it can be an asset in the hands of the spinners. A separate chapter is devoted to Industry 4.0 initiatives to be followed up by spinning mills to become smart manufacturers.

In the chapter on "Quality and Process Management in a Spinning Mill," the author covers all the quality management practices, starting from raw material inspection to final-process inspection. Today, mills have a lot of customer complaints on "contamination" issues in final fabric/garment from the end user. Estimation of contamination index in cotton is explained, which will help spinners to decide on the measures to be taken for reduction of the same during the process. The key performance indicators (KPI) in each process have been discussed in depth and suitable actions to maintain the parameters under control have been recommended. Statistical tools that are necessary for quality assurance management in spinning mills have been covered in detail. Norms based on industry practices for incoming material quality, in-process quality, and final-process quality are highlighted for critical-quality characteristics. Classimat faults and package defects are explained in simple terms. Methods to conduct process audit and quality audit have been explained with structured formats.

The author has covered "Maintenance Management in Spinning Mills" in Chapter 2 detailing how good machinery maintenance contributes toward high productivity and quality. Neglect of maintenance and the timely replacement of parts/critical components will prove several times more expensive compared with the costs associated with a proper maintenance programme. In this executive handbook, the author covered the important elements of maintenance with a focus on spinners who deal with premium products and who wish to contribute toward excellence in manufacturing. Types of maintenance starting from planned maintenance to today's smart prescriptive maintenance framework (SPMF) have been covered in detail. The maintenance concepts like TPM/5S and FMEA also have been explained in detail. Model checklists for online and offline machinery maintenance audits in spinning mills have been provided.

Chapter 3 explains the process of effective production management to ensure the product meets the end-use performance characteristics and satisfies customers. Critical KPI's like utilization efficiency, production efficiency, and yarn realization have been discussed in detail with the measures to be followed by the spinners to excel in their contribution. Manufacturing requirements for sustainable products are also briefly covered. Model spin plan and model mill-standard documents are also provided, which are vital tools for the production planners and executives.

The author has devoted a separate chapter on "training and development" as it is an essential branch of successful management systems for sustaining the improvements and establishing continuous improvements. Training is not limited to productivity alone but it covers safety at work, personnel motivation, and personnel development. Organizations that are conscious of the role of training invest their focused time-bound plans with a flexible budget. The money invested on training and development of employees will certainly pay back if one needs to deal with it commercially. During the period of Industrial Revolution, Taylor explained the theory of scientific management. He explained about the importance of training for high productivity, low accident rate, low wastage, and maximization of profit. Standards for training the shop-floor operatives in critical departments and also the evaluation criteria also have been explained in a scientific manner with structured formats. Part analysis training (PAT), which is a disciplined system of teaching the employees "skills" compared to conventional "unsystematic training," has been explained. Management development programme has been explained in detail. Over and above, the HPT concept, which is a set of methods and processes for solving problems or realizing the opportunities related to the performance of the people, has also been covered.

The chapter on "energy management" covers the necessity of keeping the energy cost at a lower level to reduce the overall yarn manufacturing cost. Measures are explained in detail to track the energy consumption in every segment of the manufacturing operations and take innovative steps to control and bring it to the lowest level possible and also take measures to benchmark our mills against the industry standards. Energy consumption in a spinning mill and the recommended norms for energy consumption has been explained in detail with examples. Innovative energy conservation projects as implemented in mills have been highlighted for the benefit of all the spinning mills. Quality of power is more important, as it decides the efficiency of the motors and other electrical devices including power electronics components. One of the critical factors affecting power quality is "harmonics" and, hence, it is dealt in detail on control measures with suitable filters to save the electrical devices and also to reduce energy consumption. Similar to the machinery maintenance audit in spinning mills, "energy audit" is also explained in detail so that the mills can have an effective energy management. Developments in motor technology have been explained and the losses due to the use of rewound motors also have been explained. Renewable power generation is in increasing trend with a global vision of "net zero by 2050." The need for wind energy and solar energy for the spinning mills is also touched upon briefly.

In Chapter 6 on "customer focus," the author explains how to make an organization customer-focused to excel in customer relations management and also the measures essential to delight the customers. Customers for the spinning mills have been dealt in detail under different categories based on end-use requirements. Expected requirements for these customers have been explained in an elaborate manner by the author. Achieving customer satisfaction is the primary objective of any manufacturer and, hence, the procedures for achieving the same are explained with the help of customer satisfaction audits and also through the "estimation of customer satisfaction index (CSI). Methods to achieve customer delight and customer retention are explained through co-creation and taylor-made products.

A separate chapter is devoted to "Industry 4.0," which should be a vision and journey for every spinning mill. Organizations implement Industry 4.0 initiatives and prepare to turn the clearly documented Industry 4.0 vision, concepts, principles, technologies, and architecture into reality within their context. Germany launched the project "Industrie 4.0" to digitize manufacturing in 2011 at Hannover. Moving beyond its roots, today there is a global transformation toward digital transformation of manufacturing and other industries. Evolution of Industry 4.0 has been chronologically explained starting from the "First Industry Revolution 1784." The salient features of cloud computing system/Big data/IoT have been explained. Smart manufacturing with the help of cyber tools and also the capture of Industry 4.0 initiatives by the machinery manufacturers have also been elaborated. Examples are the systems that are making inroads in spinning mills, like individual spindle monitoring systems, production monitoring systems, health monitors, energy management systems, and e-bikes with dashboard. Excellence in spinning mills calls for smart manufacturing by working toward benchmarking and scaling themselves up toward newer limits.

All the chapters are provided with proven and practical case studies by the author, which add a strong value for the book by increasing the confidence level of the readers. It is hoped that the modest contributions made through this book will definitely contribute toward manufacturing excellence in spinning mills by enhancing the contribution from the readers.

A. Kanthimathinathan

Acknowledgments

The author would like to express his sincere thanks and gratitude to all the academic experts and industry leaders who reviewed the manuscript and enriched it through their value-added feedback. Author likes to place it on record his hearty thanks to the entire team of Taylor and Francis (CRC Press) under the guidance of Shri Gagandeep Singh for their enthusiastic support and guidance during the preparation of the book.

The author has immense pleasure in acknowledging the acceptance of Shri Ashwin Chandran, CMD, Precot Meridian Limited and Chairman, SIMA to glorify this book with his valuable foreword.

The author would also like to register his whole-hearted gratitude and thanks to the managements of the following spinning mills, who have given consent to use the results of the case studies and system improvements achieved by him during the tenure of his consultancy services:

1. M/S Sri Lakshmi kantha Spinners, Sadhasivapet, Telengana, India
2. M/S Sri Saravana Mills Pvt Limited-Unit 1&2, Dindigul, Tamil Nadu, India
3. M/S Vasantha Spinners Limited, Thimmapuram, Andhra Pradesh, India
4. M/S Thenpandian Spinning Mills India Private Limited, Nambiyur, Tamil Nadu, India

The author would also like to express his gratitude to his family members who have been the source of encouragement and inspiration by providing support at the various stages of bringing out this book.

A. Kanthimathinathan

Author biography

A. Kanthimathinathan, is a management consultant for the spinning mills through his own outfit WINSYS SMC, operating from Coimbatore, India. He is a post-graduate in Textile Technology from PSG College of Technology, Coimbatore, and later acquired PG Diploma in personnel management and industrial relations from NIPM, Calcutta. He has 42 years of experience in reputed spinning mills and in a research association. He has more than 30 technical papers published in industry and research journals. One of the book, co-authored by him, won the "Best Book Award" from the Textile Association of India. He is also one of the recipient of National Research Development Council (NRDC) Award for a joint contribution toward an UNDP-sponsored project. He contributed for nine patents for product development while serving at The South India Textile Research Association (SITRA), Coimbatore. He had been deputed to Leeds University, United Kingdom, and Clemson University, United States of America, by SITRA on an UNDP-sponsored project to study and acquire knowledge from academic experts. He had visited many mills in India and few mills abroad on technical consultancy visits through SITRA as well as his through his own consultancy organization. He shared his technical knowledge through various platforms like national-level textile conferences, seminars conducted by research associations, textile associations and academic institutions, etc. As a CEO, WINSYS SMC, he has conducted several technical seminars across India for the benefit of professionals serving in textile industries.

1

Quality and Process Management in a Spinning Mill

LEARNING GOALS

- Operations in a Spinning mill
- Incoming Material Quality Evaluation
- Contamination Index
- In-Process Material Evaluation
- Package Defects and Control measures
- Application of Statistics
- Final Inspection of Products
- Key Performance Parameters in Each Department
- DUO Clearers

Introduction

Spinning mills in India have a lot of challenges both internally as well as externally. Even though trade agreements are affecting us in one manner, the other side is the efforts from mills for meeting Export yarn Quality standards– consistently. Towards this end SITRA has recommended few steps as below:

- Acceptance of Export culture through company wide communication and follow up
- Proper training of Quality control to all the concerned employees
- A well equipped laboratory and a full fledged Quality control department
- To impart training to operatives for avoiding bad work practices
- Strict inspection procedures & standards to avoid passing of defective products to the customers

DOI: 10.1201/9781003257677-1

Quality has many definitions. In whichever manner we define it, it is clearly understood that we need to meet the "Agreed/Perceived level of Quality level "for any products we manufacture. It is better to understand much more important definitions for "Quality"

- **Quality** is the totality of features and characteristics of a product or service that bear on its ability to satisfy given needs. (***American Society for Quality***) Quality, an inherent or distinguishing characteristic, is a degree or grade of excellence.
- **Quality** can be defined as conformance to specifications.
- **Quality**: more than asked for, more than expected, beyond common maximum perceived value, for free.
- Perhaps the most useful ***Business definition*** of quality is "fit for purpose." This definition evolved in Quality management circles. It's useful because it's applicable to any process, service or product.
- Economists tend to judge quality by the price consumers are willing to pay.
- Quality is "compliance to the best known standards, processes and specifications."
- Quality is "a satisfying experience-Customer's delight"
- W. Edwards **Deming defined quality** as follows:

Good Quality means a predictable degree of uniformity and dependability with a Quality standard suited to the customer. *The underlying philosophy of all definitions is the same – consistency of conformance and performance, and keeping the customer in mind.*

In a nutshell, Quality means "conformance to agreed standards/customer requirements in a consistent manner at an economical cost." Organization should establish its Quality Assurance department has the required systems to ensure that this objective is achieved by effective utilization of resources.

1.1 Spinning Mills-Flow Chart of Operations

Normally in any Spinning mills, the regular flow of operations will be as given in the sketch shown below:

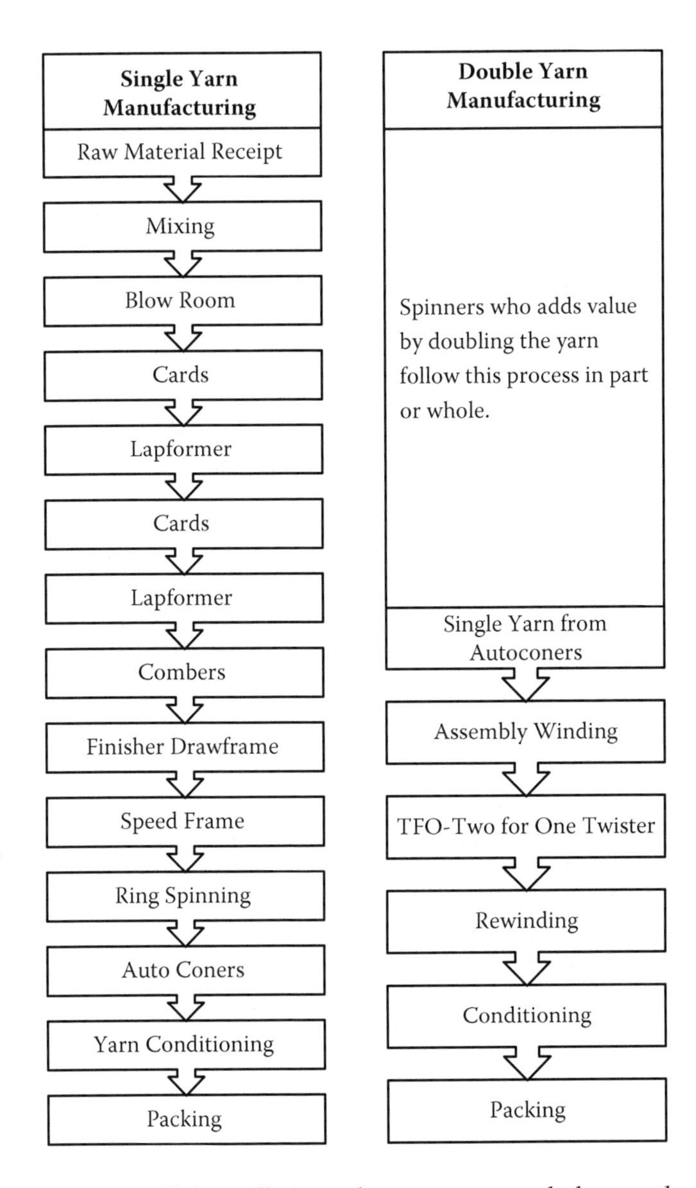

Note 1: If the mills intends to process carded yarn, the flow of material will be Pre comber Draw frame to Finisher Draw frame and continue. In that case we will be calling the Pre Comber Draw frame as "Breaker Draw frame."

Note 2: If the mills process Compact yarn, Ring frame will have Compact attachment.

Mills should have a structured Organogram in QA department (Annexure I) to manage the Quality Assurance Schedules (Annexure II) and the models are provided in Annexure. Head of QAD should organize these scheduled activities and Special studies towards meeting the customer requirements.

1.2 Quality of Incoming Raw Material (Cotton)

Cotton, raw material, normally is received at mills in the form of Bales and the weighment ranges from 160 kilograms to 290 kilograms depending on the station from where it is dispatched.

Once the cotton bales are received at mills we need to test all these Quality specifications and compare against our contracts:

I. Moisture content
II. Trash content
III. Fiber length characteristics/Fineness/Maturity/Strength/Short fiber content/Color characteristics (Rd and +b) and neps

Moisture Content

Moisture content in bales is an important techno commercial parameter. The standard moisture content for cotton is 7.5 % plus or minus 0.5. If the moisture of the incoming bales is higher than the stipulated norms, Spinners are bound to lose in money terms equivalent to the extent in which it is deviated. It is vice versa if the moisture content is decreased. However higher moisture, technically leads to a lot of operational issues like sticking tendency/poor waste liberation/low Yarn realization etc. If the Moisture content is lesser to standard, in all probability, it leads to fiber rupture in Blow room and Carding and also leads to higher yarn hairiness.

With the help of a Moisture measurement meter we should test 3 samples from each bale for five randomly selected bales per lot received at our end. If the results deviate from the standards, it is always better to recheck from different set of five bales in the same manner. While drawing samples from each bale, one should take care we measure the top/middle/bottom portion of the bales without fail.

Trash Content

Trash content of the bales need to be tested for at least one bale per lot. If the lots are produced and dispatched from the same ginner, we can relax our schedule based on the consistency of the results. Trash content needs to be tested by using a Trash separator instrument developed by Research Associations and manufactured by several testing equipment suppliers. Latest equipments has the provision of assessing the level of micro-dust also along with the trash content.

The test results can be compared against Industry norms as below:

TABLE 1.1

Level of Trash content

Type	Very good	Good	Average	Poor
Indian cotton	<1.5	1.5–2.0	2.0–3.5	>3.5
Imported Superfine cotton	<1.2	1.2–1.8	1.8–2.5	>2.5

Mills should have the practice of claiming from raw material vendors if the trash content exceeds the standard level specified in purchase terms suitably.

Length, Fineness, and Strength Characteristics

Mills can use High Volume Instruments for Length/strength /Fineness/ Rd/+b/SFI. All lots need to be tested and recorded. Under lot management mills can check 5% of the bales /lot upto a lot having 100 bales and 4% beyond a lot size of more than 100. For example, if the lot size is 50 bales we need to test 3 bales (2.5 rounded off to 3) and 5 bales for any lot having 100 bales. In each bale tufts need to be collected from all parts of the bale so that it represents the true (nearest) quality of the fibers in it. For each bale we need to have five specimen tests and the total average will be the "Incoming raw material quality for the particular lot."

In HVI instrument mills follow normally ICC mode only even though HVI mode of testing is followed by foreign countries for imported cotton and very recently research associations also are promoting HVI concept. In this handbook we deal only with ICC mode of evaluation and any reference on Raw material quality specification is based on ICC mode only, to eliminate any confusion in the minds of the readers/users.

ICC Mode of Evaluation

Using HVI or equivalent instruments the following characteristics need to be tested.

- 2.5% Span length/50% Span length
- Uniformity Ratio(Ratio of 50% SL to 2.5% SL)

- Fiber fineness(Micrograms per inch)
- Fiber Maturity
- Fiber strength(gm/tex)
- Short fiber index
- Color grade(Rd/plus b values)
- Trash grade

(Mills normally use HVI or its equivalent instruments mainly for the above characteristics assessment and Bale management except for trash. For Trash content measurement, SITRA trash separator is considerably the best instrument which correlates well with the process fine tuning system.)

AFIS instrument is used to estimate the fiber neps in each lot in the same samples taken for HVI. AFIS instrument provides results on Nep size and neps per gram.

1.3 Mixing Quality

Once the raw material specifications are arrived at, the Mixing issue chart for a minimum of fifteen days should be prepared and communicated to the Production department to lay the mixing as per the chart. Depending on the end-use of the product we should take care of "range" of fineness and length in the mixing. For hosiery products fineness should be maintained within a range of 0.2 whereas it can be upto 0.6 in the case of Weaving counts. As far as the length is concerned it is recommended not to mix Cottons varying beyond 1.5 mm in a mixing as roller settings in any drafting system will be very difficult to optimize. But most of the time it is not possible to meet this requirement.

Mixing issue chart needs to be prepared with other considerations as below:

- Less trashy cotton if it is meant for compact counts as dust level affects the quality of compacting in drafting zone
- Less or nil contaminated cotton if it is meant for Hosiery counts
- Less variation in fineness of cotton if it is meant for piece dyed knitted fabrics
- Lots with higher maturity coefficient if the end use is premium knitted products as less mature cottons will lead to white neps

Based on the fiber quality specifications FQI is estimated. CSP(Count Strength Product) of a yarn,which is an important Yarn quality characteristic can be predicted using FQI. This prediction formula has been developed by SITRA and it is being reviewed periodically too.

FQI &CSP Prediction (as per SITRA)

FQI-Fiber Quality Index: A single measure as established by SITRA for the quality of the cotton is FQI-Fiber Quality Index.

$$FQI = Lsm/f$$

Where L represents 50% span length/s represents fiber strength in gm per tex/m stands for maturity coefficient and f represents fineness of the cotton in Micrograms per inch.

Prediction of CSP for Carded Counts:

$$\text{CSP(Count Strength Product)} = K * \sqrt{(Ls/f)} + 590 - 13\,C$$

(K = Constant of 285 and C represents the actual spin count of yarn)

Prediction of CSP for Combed Counts:

$$\text{CSP(Count Strength Product)} = \{K * \sqrt{(Ls/f)} + 590 - 13\,C\} * (1 + W/100)$$

Where "W" represents Comber noil extraction in percentage.

Mills can lay the mixing as per the CSP required in the final product and the process parameter in each machine needs to be decided based on fiber quality and end use requirements. Quick Sample evaluation for fixing process parameters in each department can be done if the Cotton is of new variety or new arrivals of the cotton season.

A special word about contamination level in the yarn needs to be mentioned before proceeding to the Blow Room. Cotton, as it is manually or mechanically picked from the plants and transported to the ginneries there are enough opportunities for getting contaminated with colour yarns, colour fabrics, feathers, Polypropylene/polyethylene twines or sheets apart from full seeds/half seeds/leaves inherent to cotton. Contamination index needs to be assigned for each lot by assessing randomly selected bales from each lot and manual segregation and counting all these contamination. A simple method of Estimation of contamination index is suggested in Annexure III.

1.4 In-Process Quality

Quality assurance department should ensure the "Quality is built into the product" for which materials "in-process" must be evaluated at every department.

1.4.1 Blow Room

OBJECTIVE OF BLOW ROOM

Cotton is transported to the mills in compressed bales of 170 kg, 220 kg, 245 kg and 290 kg. Indian bales normally weigh only 170 kg. All imported bales are with a high level of Bale density to reduce transportation cost from abroad. It requires "opening of bales into smaller flocks."

Apart from that, cotton being naturally grown and picked from the fields (manually as well as mechanically) has a lot of natural trashy elements like full seeds/broken seeds/leaves/stems/sand dust/small stones etc. All these vegetable matters and heavy trash particles need to be separated out effectively in the Blow room once the flocks are opened.

Over and above purpose of Blow room is to ensure homogenous mixing of different varieties or different lots of the same varieties.

Blow room is expected to open and clean the fibers and deliver it to the next machines i.e. Cards with an acceptable quality. Cards need to be given more or less as a cleaned cotton,which is well opened.

Hence the cleaning efficiency as well as the opening efficiency of each beater needs to be assessed by Trash separator and Opening efficiency tester. However, once use of an Opening tester is not popular in the industry mainly because of the inconsistency in results and "hand feel" judgement overrules the instrument usage.

Cleaning Efficiency with Trash Separator

Modern Blow room has several Openers/cleaners/Mixers and Beaters as below:

- Bale pluckers/Blendomats
- Uniclean/Vario clean/CL-P precleaner
- Unimix with beater and without beater/Multimixer
- ERM Cleaner/Cleanomat (1&3)

We need to evaluate the cleaning efficiency of each Cleaner and Beater by assessing the trash contents in input material and output material. Recommended sample size is 100 grams and no of tests should be minimum 2 while we set the process. Otherwise during the regular process it is enough to test one sample.

$$\text{Cleaning efficiency} = \{(\text{Trash content in input} - \text{Trash content in output}) / \text{Trash content in input})\} * 100$$

As per the industry norms, we should achieve the following results for cleaning efficiency:

First stage cleaners : 35 to 40%
Second stage cleaners : 20 to 25%
Third stage cleaners : 10 to 15%

The overall cleaning efficiency of the Blow room should be well above 65% for cottons with a trash content above 3.5% and 50–55% for cottons with a trash content of 2–3%.

For imported cotton with very less trash we need to concentrate on opening efficiency only and cleaning efficiency can be around 40 to 45%. Moreover imported cottons are being received with higher bale density (to reduce transportation cost) and hence we need to focus on the quality of opening in the Blow room.

Control of Neps in Blow Room

Neps are small tiny knot like aggregates of entangled fibers with or without fragmented seed coat/leaves. They show up predominantly in yarn and account for a large proportion of customer complaints received by mills manufacturing premium products. High incidence of neps is responsible for poor fabric appearance and appeal characteristics of yarns and fabrics, formation of spotty, streaky materials during dyeing and printing, end breaks during Spinning, Clearer cuts in Autoconer winding, Breaks during Warping and weft insertion in Looms, especially in Airjet looms & in knitting.

Mixing will contain neps as an average of the constituent lots. Neps tend to increase by the "air turbulence and beating action" in cleaners/beaters. The surface quality of the Cleaner/Beater pins & hoods and Conveying suction ducts and suction quality also are prime factors deciding on the level of neps generated during the Blow room process.

Neps during in process is being assessed with AFIS instrument and random samples from the in process material needs to be collected similar to estimation of Cleaning efficiency. However the quantity of sample required is less only but care must be taken to make it representative.

In any kind of Blow room there will be a tendency for nep generation and we can only contain the level of generation and it cannot be eliminated. As an industry standard, we recommend control standards as below:

TABLE 1.2

Nep Generation Level in Blow Room

Factor	Good	Average	Poor
Nep Generation %	**80–100**	100–130	>130

1.4.2 Cards

OBJECTIVES OF THE CARD

Main objective of the cards is to "individualize" the fibers from the clean "cotton bait" supplied to it by Blow room. During the process it eliminates almost all trash particles –without affecting or with least impact on fiber length. Short fibers associated with trashy particles are eliminated in the Carding process.

Cards are playing a vital role in deciding the final product quality. Hence modern cards are equipped with precise Auto Levellers, Pre and Post carding segments and well designed wire points for Cylinder/Flat tops and Doffer. Moreover the Active carding surface is focused by all the leading manufacturers of Cards today so that the assignment given to cards is well done.

In Cards our focus will be the Control of sliver hank and its mass variation, Cleaning efficiency and Nep removal efficiency. Sliver hank can be maintained by ensuring uniform feeding to cards, proper tuft opening in Blow room which involves suction level controls in Blow room and Chutes.

Mass variation in Sliver can be controlled by precise setting of Autolevellers/ regular cleaning and calibration of Autolevellers & its mechanical parts.

Nep removal efficiency depends on:

- Card waste extraction level (Recommended level: 2.3*Trash content in mixing)
- Condition of wire points
- Settings in vulnerable zones i.e. Lickerin to feed plate/Lickerin to Moteknife/Pre and Post carding cleaner settings/Trash knife settings and Flat tops to Cylinder settings.

Mills must assess the level of waste generated and also the "quality of the waste generated in each zone." As per the industry norms we must try to achieve NRE of 75 plus always. Any NRE level below 65% calls for reviewing input trash level/Wire condition in Cards/Critical settings in cards and waste generation.

Fiber rupture during the Carding process can be estimated by estimating the SFC(n) of the input material and output sliver. As the cards remove short fibers associated with trashy material, we can expect an improvement in SFC(n), But due to the severity of carding action of fiber individualization fiber damage is unavoidable. However the extent of deterioration in SFC(n) in output sliver should be restricted within 6%. If it exceeds the limit we should fine tune the Speeds and settings in Critical zones.

1.4.3 Pre-Comber Draw Frames and Lap Formers

OBJECTIVE OF PRE COMBER DRAW FRAME

PCD needs to double and improve the evenness of the Card sliver by proper drafting system.

OBJECTIVE OF THE LAP FORMER

Lap former is expected to further double the doubled and drawn slivers for improving the evenness of the bait and also to prepare a lap suitable to be fed to the Combers with uniform distribution of ***"no of fibers per unit length-in width and length."***

In Pre-comber Draw frames Sliver evenness must be fine tuned below 3% and the number of doublings should be 5 to 6 only. In Pre-comber Draw frames, the bottom roller conditions and its saddles must be well maintained by assessing the U% and its variations.

In Lap formers, the number of Doublings is decided by the Lap weight in grams per meter and also with due consideration on lap licking tendency. Higher the draft in Lap former, higher will be the licking tendency. As the modern lap formers are capable of producing lap weight upto 25 kg, the inverter controls must be properly set for achieving less mass variation between initial and final layer of the laps produced.(CV% of hank should be below 1.5%).

1.4.4 Combers

OBJECTIVE OF COMBERS

Combers need to remove the short fibers in the lap/straighten the fibers individually, double it, draft it and coiled into an even/clean Combed sliver. In this process of Combing we also aim at reduction of neps and thick places. Combing process is a part of value addition for Quality conscious customers. By suitably designed Comber noil extraction, final yarn quality can be fine tuned to meet customer's requirement and end use performance characteristics.

Combers contribute a lot for the improvement of final product in terms of

- Aesthetic appeal
- Yarn imperfections
- Yarn strength
- Classimat faults
- Hairiness

In Combers all the fibers are parallelized during the short fiber extraction process with the best use of Top combs/Cylinder/Brush in combination. Moreover as the modern Combers has single delivery for all the eight laps fed with a sturdy Drafting system, evenness of the delivered sliver is ensured. Combers should be set for the required nep removal efficiency, optimum Comber Noil extraction and the best possible U% for enhancing the final yarn quality.

Comber Noil % must be set within a tolerance limit of 0.5%. At periodic intervals CN must be tested and corrective action must be initiated. Head to head variation also must be controlled within 1% range in modern combers. While assessing head to head variation, manufacturers' recommended methods must be followed. Moreover, if the combers are connected to separate compactors for each mixing "bulk level studies also can be done" whenever there is a mixing change or once in a month.

Efficiency of the Combers are being assessed with two important critical factors:

- Nep removal efficiency(depends on the noil extraction)
- Short fiber removal efficiency

Nep removal efficiency is estimated in the same manner as we follow in Cards. Nep removal efficiency depends mainly on the level of Noil extraction. QA department needs to check on the level of neps periodically for the same level of noil extraction. Normally mills maintain 65% and above for Premium quality customers.

Short fiber removal efficiency should be estimated as below:

$$= \{(\text{SFC(n) in the Comber lap} - \text{SFC(n) in sliver}) / \text{SFC(n) in Comber lap}\} * 100$$

Short fiber removal efficiency also depends on Noil extraction level. As per the industry practice mills maintain the following targets for SFC(n) in Comber sliver.

TABLE 1.3

SFC(n) in Comber Sliver for Premium Counts

Count range	Premium Hosiery counts(less than)	Premium Warp counts(less than)
20s to 34s	12	11
36s to 40s	11	10
44s to 50s	10	9
60s to 74s	9	8
80s	7	5–6
100s–120s	–	4–5

Importance of SFC(n) needs to be well understood as it has an impact on Ring frame end breaks/Autoconer clearer cuts/Classimat faults and Warping breaks. Hence mills producing yarns meant for Premium customers focus on monitoring and controlling the quality of in-process materials with stringent target limits.

Moreover mills must maintain Comber sliver Unevenness below 3.5% in third generation combers and aim at 3% in fourth generation Combers (like Rieter E80/LMW"s LK69). Comber sliver unevenness is affected by Piecing index, staggering of piecing in table, settings in drafting zone and the condition of rollers and its drives including the weighting system.

1.4.5 Finisher Draw Frames

> ***OBJECTIVES OF THE FINISHER DRAW FRAME***
>
> *Finisher Draw frame is the final preparatory process in the sense that quality of the product cannot be improved beyond this process, but only can be deteriorated. Purpose of the Draw frame is to double and draft the optimum number of fed slivers to make it more uniform with least mass variation with the help of Autolevellers.*

Draw frame is the final preparatory process beyond which no more efforts can be initiated to improve the quality of in-process further. It can only be affected negatively by the Speed frame process. Hence, the Draw frame process is more critical for final product quality. The objective of Drawing process is to "produce more even slivers" by doubling process. Hence focus must be given to "Sliver Unevenness –U%" and "mass CV%."

TABLE 1.4

Industry Standards for Sliver U% in Modern Drawframes

Rating	Sliver Hank from 0.1 to 0.16	Sliver Hank from 0.17 to 0.22
Very Good	1.7	1.9
Good	1.8	2.0
Average	2.1	2.3

Premium customers need to maintain Sliver evenness at "Good" rating and if the trend shows it is shifting towards "Average" immediate actions must be initiated to restore the sliver quality.

Mass variations in Sliver at Draw Frame directly affects Count CV of yarn which has a major impact on fabric appearance. Hence mass CV of 1 metre cut length needs to be maintained.

TABLE 1.5

CV of 1 m Sliver in Modern Drawframes

Rating	CV(1m)
Very good	**<0.4**
Good	<0.5
Poor	>0.7

We have not indicated any average level here as any deviations beyond 0.7% will have serious implications in end use performance of the final product. Hence it is recommended to monitor the trend of 1 m CV % with the help of Quality Monitor in the modern Draw Frames.

Another important Quality assurance activity in Drawframe is to check A% from the Slivers. A% helps the mills to validate the performance of the Auto Levellers in Drawframes. A% Deviation helps mills to ensure the function of Auto Levellers by timely resetting and calibration.

1.4.6 Speed Frames

OBJECTIVE OF THE SPEED FRAMES

Speed frame is earlier known as "Fly frame or Simplex" too. As the speed of the machine has been considerably increased over these years(from the level of 800–850 rpm to 1400–1500 rpm) for Simplex, Industry refers to it as "Speed frames" nowadays.

Speed frames are expected to produce even roving and wound it into a convenient package of optimum weight with less mass variation and high level of unevenness. Material handling at the creel and at flyer zone is more critical and any deviations in these areas will damage all the efforts taken during preparing the Draw frame slivers.

Speed frame is a "delicate machine in Ring spinning process" where material handling plays a major role in defining the quality of the final product. Speed frames are expected to produce a package of roving with necessary twist without affecting the quality of the sliver supplied."

Three important Quality attributes are to be focused in Speed frames as indicated below.

- Roving evenness
- Roving mass variation of 3 m cut length
- Stretch %(a measure of tension difference)

Roving evenness depends mainly on input sliver evenness, draft in speed frames, quality of drafting system elements and roller settings. ndustry norms for the Roving evenness is given below:

TABLE 1.6

Roving Unevenness in Modern Speed Frames for Combed Counts

Rating	Roving hank of 0.70 to 1.5	Roving hank above 1.6
Good	<3.2	<3.4
Average	3.3-3.5	3.5-3.6
Poor	>3.8	>4.0

Roving Mass variation at a cut length of 3 metres has significant impact on Final yarn Count CV and industry norms for the same are given below:

TABLE 1.7

Roving 3m CV%

Rating	Targets
Very Good	**<1.1**
Good	1.1-1.3
Poor	>1.6

Third important quality attribute for Speed Frames is "Stretch %" and the formula for the same is given below:

$$\text{Stretch } \% = \{(\text{Difference between hanks of initial and final bobbin} \\ /\text{average of both the hanks of initial and final})\} * 100$$

In modern Speed frames, as the tension is controlled electronically, Industry norms are made stringent too. Mills need to maintain 0.3% and below for a better yarn quality and anything above 0.6% needs to be rectified.

(hank of final and initial bobbins are collected at the time of Doff/ immediately after the doff & the sample size should be twenty bobbins for each.)

1.4.7 Ring Frames

OBJECTIVES OF THE RING FRAMES

The objective of the Ring frames is to produce Normal Single yarn from the given roving, in small packages called "cops" with acceptable level of Yarn specifications to meet the Customer/market requirements, economically and ergonomically.

In Ring frames, we feed the roving bobbins and produce yarns meeting our market/customer requirements. Major quality attributes of Yarn at RF stage are:

- Count and Count CV of yarn
- Lea strength and its CV
- Lea CSP(Count strength product)
- Twist and CV of twist
- Yarn Evenness and short term imperfections
- Yarn Hairiness in Compact yarn
- Single yarn strength/Elongation/Weak places
- Snarling tendency for HT and Elitwist yarn

Count and Count CV of Yarn

Mills normally specify "nominal count" and "actual count" in practice. Nominal count represents the commercial representative number for the count of yarn to be spun and the actual count is "the count of yarn that will fit into customersend-use requirements". For example, 60s CW is a commercial representation and the nominal count becomes 60. But in practice Customer would have specified the actual count of yarn should be 61s to meet his fabric production requirement. Hence in this case the actual count becomes "61"s. All entries for instruments and production estimation only actual count needs to be utilized without fail as otherwise the result will be erroneous.

Checking the count in Ring frame is only for ensuring its consistency between days in mills with Modern Draw Frames with Auto Levelers. Count of yarn needs to be checked at periodic intervals (say once in a week by covering all the Ring Frames for each count). Fine tuning of Sliver hank needs to be done at Drawframe if the actual count is finer or coarser.

Count is estimated by preparing the leas from the cops collected from the Ringframes. Once the leas are prepared Lea strength is estimated using Lea tester. Once the strength is evaluated,with a suitable software programme we will be able to assess the extent of Variation in Actual count and Strength.

TABLE 1.8

Count CV/Strength CV(Lea strength) of Yarn

Rating	Count Range	Count CV	Strength CV
Very good	**Upto 40s**	**1.0**	**3.5**
	50s and above	**1.2**	**4.0**
Good	Upto 40s	1.1	3.8
	50s and above	1.4	4.5
Average	Upto 40s	1.4	4.7
	50s and above	1.5	5.2

Lea CSP is estimated by the actual count multiplied by its respective strength in pounds. Lea CSP is undoubtedly an established commercial parameter even today in Indian Yarn Market whether it is hosiery or Modern Airjet weaving & Industry has fixed targets for CSP for a better performance as below:

TABLE 1.9

CSP-Industry Norms

Count	Minimum level of acceptance
Upto 30s Combed yarn for piece dyed knitted fabric	2350
Upto 30s Combed Yarn for Yarn dyed process	2550
40s to 60s Combed yarn for piece dyed knitted fabric	2450
40s to 60s Combed yarn for Yarn dyed knitted fabric	2650
Upto 30s Combed weaving yarn for piece dyed woven fabric	2600
Upto 30s Combed yarn for Yarn dyed woven fabric	2800
40s to 60s Combed yarn for Piece dyed woven fabric	2750
40s to 60s Combed yarn for Yarn dyed woven fabric	2950

Twist in Single Yarn and its CV

The level of twist to be inserted in a single yarn depends on the end use of the yarn. The following table provides different twist multipliers inserted in Single yarn:

TABLE 1.10

Twist and Twist CV for Yarns meant for different end use.

Count range	End use	TM range*	Twist CV
<34s Cbd	Knits	3.55 to 3.7	2.5
36s Cbd to 60s Cbd	Knits	3.45 to 3.55	3.0
<30s Cbd	Wvg	3.9 to 4.1	2.5
32s Cbd to 80s Cbd	Wvg	4.0 to 4.15	3.0

Note: In the case of "Under spinning" or for Imported cotton mixing, TM levels can be suitably reduced by upto an extent of 8%.

Yarn Unevenness and Imperfections

The Yarn Unevenness and Imperfections form the basic Quality requirements of yarn which has commercial implications for a Spinner. Hence it is essential for the spinner to achieve Yarn evenness and imperfections to the acceptable level by fine tuning the process parameters. Even though there are Uster Quality levels in 5%, 25% and 50% for imperfections, mills follow their own norms based on "Customer's specific requirements", which need to be met by process fine tuning.

The industry norms for yarn evenness and Imperfection levels by the manufacturers of Premium yarns are highlighted as a range in the below table.

TABLE 1.11

Final Cone Yarn Evenness and Imperfection Levels in Premium Products*

Yarn type and Count	Very Good		Good		Average	
	U%<	Total imps	U%<	Total imps	U%<	Total imps
30s Combed	**9.0**	**30**	9.6	55	10.3	90
40s Combed	**9.5**	**45**	10.2	75	10.8	110
50s Combed	**9.8**	**60**	10.5	85	11.0	120
60s Combed	**10.5**	**75**	11.0	100	11.5	140
80s Combed	**10.8**	**120**	11.5	180	11.8	270
100s Combed	**11.5**	**160**	11.9	260	12.5	350
120s Combed	**12.3**	**220**	12.8	360	13.3	500

Note: Industry tries to achieve these results in premium products either by selection of suitable raw materials/fine tuning the process with waste levels/compact spinning systems.

Note: We have not covered Hosiery yarn in the 100s and 120s as they are scarcely produced.

Single Yarn Strength/Elongation/Weak Places

Elsewhere in this section we discussed about Lea CSP. Lea tester helps to find out the "bulk strength" of the yarn whereas, we need "individual strand strength" in critical end use applications of yarn. Hence, we have Single yarn Strength measuring instruments and the most popular instruments are Tensorapid, Tensomax and Tensojet. Only in the latter instrument we can estimate weak places in yarn and Tensojet results correlate well with the yarn performance in Warping and Weaving.

With these instruments, we will be able to estimate the Single yarn strength in grams, Elongation %, RKm and CV of these attributes and weak places. Mills mainly focus on RKm for all yarns meant for Weaving enduse.

TABLE 1.12

Targets for Tensile Characteristics * for Premium Products

Count group	RKm>	CV of RKm<	Elongation>	CV of Elongation<
30s Cbd weaving	18.5	6.7	5.8	7
40s Cbd Weaving	19	7	5.3	7.5
50s Cbd Weaving	19.5	7.5	5.1	8.0
60s Cbd Weaving	19.8	8	5	8.5
80s Cbd Weaving	21	8.5	4.8	9
100s Cbd Weaving	20.5	9	4.6	9.5

Note: All these targets are achievable with Indian cotton. Mills also use Imported cotton or its blends for "superior fabric aesthetic appeal" and hence the results on RKm and Elongation will differ widely. Moreover these Single yarn strength levels are enough for meeting the requirements of High speed weaving and Airjet end use and we need to focus on Package quality and weak places.

* As measured in Tensorapid testers.

Snarling Tendency

When the yarn is twisted, it has a tendency to snarl when it is suspended freely between two hands. This tendency needs to be minimized by "Yarn conditioning" since this will affect the free unwinding of yarn from the package and in Air Jet weft insertion it can cause fabric defects too. Snarl testers are available in the market for assessing the same. This instrument can be used for controlling the snarling tendency in Single yarn as well as Double yarn.

TABLE 1.13

Industry Targets for Snarling Tendency

	Single yarn and Double yarn	Elitwist yarn	High twist yarn
All counts	2.7	3.2	3.5

1.4.8 Ring Frames With Compact System

OBJECTIVES OF THE RING FRAMES WITH COMPACT SYSTEM

The objective of the Ring frames with Compact system is to produce yarns with lower hairiness level and less imperfections/higher strength/lower level of weak places & defects.The Compact system aims at higher production levels and higher yarn realization.

Compact Systems in Ring frames were introduced in the fag end of the 20th Century and now it becomes a vital need for the Spinners because of its proven results in terms of savings in Raw material/Higher production per Ring frame/Improved yarn quality in terms of Hairiness, imperfections and weak places and finally its enhanced enduse performance in Knitting and Weaving. Compact systems can be broadly classified into two systems i.e. Mechanical and Suction Compact system. In the Mechanical Compact system, there is no extra energy consumption since the system uses a mechanical compacting device in the drafting system for compacting the fiber fleece. In the Suction Compact system which is widely popular among the Spinning mills globally, energy consumption in Ring frame will be higher than the normal Ring frame by 8–12% as energy is required to develop suction for compacting the fleece.

Mills use Compact Spinning systems effectively to:

- Achieve higher yarn realization for the same yarn quality(with less hairiness)
- Achieve higher production even upto 12% for the same yarn quality(with less hairiness)
- Achieve premium yarn quality with same yarn realization and production level as in normal yarn

Hence, final yarn Hairiness levels alone are provided for compact and Normal yarn separately.

TABLE 1.14

Final Cone Yarn Hairiness levels*

Count	Very Good Hosiery	Wvg	Good Hosiery	Wvg	Average Hosiery	Wvg
30s Combed	4.5	3.8	4.8	4.3	5.5	4.8
40s Combed	3.5	3.2	3.8	3.8	4.8	4.3
50s Combed	3.5	3.0	3.6	3.3	4.4	3.6
60s Combed	3.3	2.8	3.5	3.1	4.0	3.5
80s Combed	2.9	2.5	3.2	2.8	3.7	3.1
100s Combed	–	2.3	–	2.5	–	2.7
120s Combed	–	2.1	–	2.3		2.5

Apart from the above listed benefits, Classimat faults in the Compact yarn is always significantly lower in comparison to the Normal yarn which helps to improve the aesthetic appeal of the fabric.

1.4.9 Autoconers

OBJECTIVES OF THE AUTOCONERS

The primary objective of the Autoconers is to convert cops into cones so that "transportable packages" are produced. In this process of conversion, these Autoconers clear the yarn defects and contaminations, weak places with the aid of Electronic yarn clearers, at a higher productivity level in comparison to its earlier generation machines -Conventional Cone winders." Autoconers ensure "defect free packages with a precisely controlled length per cone" through its additional optional attachments.

Autoconers are a boon to Spinners as it offers a wide range of benefits to the mills. With the right kind of Clearers to meet the requirements of the Customers Autoconers help the spinners to "Convert cops into cones, clear the random occurring-disturbing faults, eliminates Weak places, eliminates Contamination in final yarn and produces defect free packages with least length variation between cones produced."

In Autoconers cops are converted into cones with an adequate tension applied on single yarn. The purpose of this tension application is twofold.

One objective is to eliminate weak places in yarn and another purpose is to get a compact package with the optimum level of density. However, the tension applied on the yarn, if it is not optimum, will have a negative effect of increasing Yarn hairiness and short term imperfections further.

Industry has set norms for controlling the level of increase in Hairiness and imperfections as below.

TABLE 1.15

Permissible Level Of Increase In Imperfections During Autoconer Winding

Count group	Normal Yarn(<)	Compact yarn(<)
Below 30s	15	5
34s to 50s	20	10
60s to 80s	30	15
Above 80s	60	20

In Autoconers, we need to focus on:

- Splice quality(Splice appearance should be good and splice strength should be above 90% of the parent yarn strength as measured by Splice testing equipment)
- Defect free packages/Package rewinding performance (Rewinding at 1100 mpm in Autoconers itself with a simulated creel of warping). In such rewinding studies we can unearth all kinds of package defects and rectify before bulk production. Package defects and its remedial measures are provided in Annexure .
- Rogue drums where critical settings/components fail and create yarn/package defects which affects its end use. For this analysis mills use "Information and reporting system" as separately provided by Clearer or Machinery manufacturers. These kinds of MIS Reports give daily reports/Weekly reports in comparison against Benchmarked levels of performance.

1.4.9.1 Electronic Yarn Clearers (EYC) in Autoconers

Clearers in Autoconers are essential attachments in Autoconers and it helps the spinners to deliver the products meeting Customer expectations by eliminating the defects to the maximum possible extent. In 1970s most of the spinners used Slub catchers in winding machines to clear the defects and they were able to clear "disturbing thick defects" which is also only partially. From the stage of such primitive clearers we now have "Duo Clearers" which works on both Capacitance and Optical principle. Due to the importance of this subject an extensive coverage on "Evolution of Clearers" is enclosed in Annexure IVA.

1.4.10 Yarn Conditioning

Yarn conditioning is being recommended to the mills to improve:

- End use performance
- Yarn realization
- Reduction in snarling tendency

Industry needs to have three or four different levels of programs for Hosiery yarn/Weaving yarn/Elitwist yarn/High twist yarn/Double yarn so as to achieve the real benefits of Yarn conditioning.

1.4.11 Packing

Spinning mills follow several methods of packing systems in their mills depending on the Customer"s requirements. The usual packing systems followed by the Spinning mills are listed below:

A. Bag packing: In this system, High density Polyethylene or Polypropylene bags are used for packing the cones. Depending on the mode of transportation and distance to the customer location, card boards are used to cover the cones at sides/bottom and top. All cones are individually put in separate polythene bags of 40 micron and above. This is one of the cheapest mode of packing available as on date. Mills use laminated or un-laminated bags depending on the necessity to safeguard from damages due to rainy season at the time of transportation.

B. Carton packing: Suitably designed Carton boxes are used to stuff a fixed quantity of full cones. Normally quantity per carton ranges from 12 to 24 and the individual cone weight ranges from 0.945 grams to 2.85 kg. Few customers demand cartons covered with HDPP bags too as a foolproof against any kind of transportation damages.

C. Pallet packing: With the help of a Pallet packing equipment, pallets are prepared with 12 or 14 layers of intermediary hole pads. These staggered layers are wrapped gently with a stretch film to tightly cover all the cones. In each layer normally mills arrange 6*6 cones or 8*8 cones depending on the cone weight specifications. The number of layers is decided based on the type of container. In the normal container we can transport only pallets with 12 layers and in Hi-cube container we can transport pallets with 14 layers. For exports and long distance transportation, customers worldwide prefer pallets only as handing of huge quantity is easier and less time consuming with the help of fork lift/battery operated trolleys. Mills also prefer this method of packing as the opportunity of damages to the cones are the least in comparison to all the other packing mode. However we need to be cautious to ensure:

- Wooden pallets used for packing is free from termites/ insects (for ensuring the same we normally do fumigation of pallets before packing the goods)
- Stretch film quality should be as per the specification of the customers and normally LLDPE with 40 microns along with 300% stretchabilty and 500 mm width is being preferred.

Packers must be well trained on Customer wise requirements and they should be trained to remove any kind of defective cones/less and overweight cones and keep it aside. Since packers are the source of our final inspection, we need to ensure everyone involved in the packing process is aware of all kinds of package defects. Hence it is advisable to keep a visual communication board with names of the defects and its corresponding photos in colour.

1.4.12 Assembly Winding and Two for One Twister

OBJECTIVES OF THE TWO FOR ONE TWISTER

The objective of the TFO-Two for one twister is to manufacture ply yarns from parallel wound single yarn packages from Assembly winder. Knitters and Weavers demand plied yarn for their enhanced fabric qualities in terms of Aesthetic appeal, drape behavior, comfort properties, tearing strength etc. In TFO we twist the single yarns into required ply yarn. Normally 2 ply/3 ply yarns are more popular in the Knitting/weaving industry.

In TFO, the following properties need to be focused:

- Single yarn strength(RKm) and weak places
- Yarn imperfections
- Twist and its variation
- Yarn hairiness
- Snarling tendency

Achievable results on all these characteristics depend on the qualities of single yarn sourced/produced. However the industry has set targets for doubled yarn expected strength from Single yarn quality levels.

TABLE 1.16

Expected improvements in Yarn imperfections & Strength after Doubling

Factor	Strength improvement	Imperfections reduction
Doubled yarn against Single yarn	>10%	>80%

1.5 Classimat Faults

Classimat faults in yarn are the random faults as generated during the process due to Raw material as well as process conditions. It is exclusively dealt in Annexure IV B.

1.6 Package Defects –end Use Performance

It is the responsibility of the organization to meet the Customer's requirements in totality by meeting his stated needs on paper and also his implied needs. Primary implied need is to ensure the "end use performance of our products." End use performance depends not only on "inbuilt quality of the yarn" but also depends on the "Quality of package." Hence we have unearthed almost all kinds of package defects that have a decisive impact on its end use performance. Considering its importance, a separate Annexure is devoted on "Package Defects" (Annexure V).

1.7 Process and Quality Audits

Process Audit is an essential activity associated with the Quality Assurance department. As a routine schedule process audits must be conducted in a structured format to find out the deviation in process parameters and every audit deviation must be closed with no time delays as Spinning mills process is a continuous one. Any deviation in process parameters will definitely affect the Quality of the product. During the process audit, Auditor needs to check on the "running process quality" through visual examination on the actual process parameters being followed against Mills's standard process parameters. Hence this Audit is called as the Process audit. During Quality Audit, Quality of the ongoing process is checked visually on a fixed interval depending on the type of product and the customer requirements. Structured formats for Process and Quality Audit are given in the Annexure VI and VII separately.

1.8 Application of Statistical Techniques in Spinning Mills

Role of a Quality Control Department is to prevent defect generation for which Statistical tools and techniques are being selectively used to assist the decision makers' function easier and more scientifically. Let us have few essential statistical techniques applied in Spinning mills as of now.

Mean

Mean or "Average" of the population is the sum of all the readings in a given test and divided by the number of readings. For example, let us assume the tested values of the Count in a Ring frame as below:

TABLE 1.17

Test results of Yarn Count in a Ring frame(Count 60sC)

58.9	58.5	59.2	61.2	**61.4**
60.6	59.8	60	61	59
58.4	61.3	61	60.8	59.2
59.3	59.5	60.6	61.1	60.9

In the above population of tested count values, the number of samples tested I,e, is "n" which is equal to 20.

Sum of all the readings = 1201.7
Number of Samples "n" = 20
Average = 1201.5/20
= 60.085

Hence Average Count in the particular RF will be concluded as 60.085s.

Range

Range is a measure which denotes the extent of the variation in a given set of readings tested at a time. It is the difference between the lowest and the highest reading in the population (set of readings)

In the above example, the highest reading is 61.4 and the lowest reading is 58.4 and hence the range is.

$$R = 61.4 - 58.4 = 3$$

This means the Count in the Ring frame can range between extreme values having a difference of 3.0

Standard Deviation and Variance

By estimating "Mean" and "Range" we cannot conclude on the behavior of the count of yarn in the particular Ring frame and hence we depend on Standard deviation and Variance measures of the distribution of the tested readings.

Variance is defined as the square of the deviations of the individual observations from the mean.

Variance of the above population is 1.021 as estimated and the standard deviation is 1.01 for the above distribution.

Standard deviation is normally denoted by "Sigma" (σ)

Standard deviation is an useful tool in the testing of materials in Spinning for making decisions as it gives a measure of the standard deviation of the tested values from the "mean."

As most of the tested values in Spinning mills follow normal distribution, the range of the tested values can be estimated using Mean and Standard deviation. In a normal distribution,

67% of the tested values will lie within Mean +/– Sigma i.e. in the above case,

60.085 + 1.01 and 60.085 – 1.01 i.e (61.095 to 59.075)

95% of the values will lie within Mean +/– 2*Sigma i.e. in the above example,

60.085 + 2*1.01 to 60.085 – 2*1.01 (62.105 to 58.065)

Likewise 99% of the values will lie between Mean +/– 3*Sigma limits ie 63.115 to 57.055

Coefficient of Variation

The coefficient of variation (CV) is the ratio of the standard deviation to the mean. The higher the coefficient of variation, the greater the level of dispersion around the mean. It is generally expressed as a percentage. The lower the value of the coefficient of variation, the more precise the estimate.

$$CV\% = (\text{Standard deviation}/\text{Mean}) * 100$$

In the above set of readings,

$$\text{Coefficient of Variation} = (1.01/60.085) * 100 \text{ i.e. } = 1.68\ \%$$

Coefficient of Variation is one of the important statistical techniques used in Spinning mills for comparison of Count, Lea strength, Single yarn strength and Yarn imperfections. Level of these variations is one of the prime factors of performance in Spinning mills.

An illustrative example for a better understanding on CV:

In a Tensorapid Strength result for a particular lot inspection of 60s Count, the Mean of Single yarn strength is 197 grams and CV of the 100 readings is 9.5, Let us try to find out the 95% confidence limits for this Single yarn strength.

As explained elsewhere, in a Normal distribution, 95% of the population will lie within Mean +/− 2 Sigma levels only(Sigma-Standard deviation) Hence we need to estimate Sigma now.

$$\begin{aligned}\text{Standard deviation } \sigma &= (CV * \text{Mean})/100\\ &= (9.5 * 197)/100\\ &= 18.72\end{aligned}$$

Hence 95% of the Single yarn strength values will fall between 197+/−2* 18.72 i.e.

159.6 to 215.72. Head of the QA department can assure the customers that his Single yarn strength will lie between these values only for the lot under inspection.

Critical Difference

CD-Critical Difference is an another important Statistical technique to assess the level of significance of the difference. It is a measure of the difference between two values that arises solely due to natural variations. When the difference between the two values crosses this CD, then we can infer that these two values "are statistically differing "or" the difference is statistically significant."

TABLE 1.18

Number of tests and Critical Difference % for various fiber properties

Fiber property	Number of tests	Critical difference(% of mean)
2.5% Span length	4 combs per sample	4
Uniformity Ratio	4 combs per sample	5
Micronaire	4 plugs per sample	6
Fiber strength(3 mm gauge)	10 breaks per sample	5
Maturity coefficient	600 fibers/sample	7
Trash content	8 tests per sample	7

Example: Fiber Rupture in Blow Room Line

For assessing the fiber rupture in the Blow room line we need to test the 2.5% Span length of the input material and the delivered material from the last beater. Let us assume:

2.5% SL in Mixing laydown = 34.5 mm
2.5% SL in the delivered material from Blow room = 32.5 mm
Difference between the two observations = 2
Mean of the observations = 33.5

Critical difference = (2/33.5) * 100
= 5.97 which is greater than CD of 4 as per above table.

Hence we can conclude that there is a significant fiber rupture in the Blow room line for this process where the sample is collected.

Similar to the fiber properties, the following table can be used for assessing the CD for yarn properties.

TABLE 1.19

No of tests and Critical Difference for Yarn properties

Yarn Quality characteristic	Number of tests	Critical difference(% of mean)
Lea count	40	2.0
Lea strength	40	4.0
Single yarn strength	100	2.8
Evenness(U%)	5	7
Twist –Single yarn	50	3.4
Twist –Double yarn	50	2

Control Charts

Control charts are used for all our "Controlling procedures of wrapping/ Count/Strength" in a more effective manner.

In control charts, the Centre line is drawn for representing the "Mean" and the 2 sigma limits are warning limits and 3 sigma limits are Control/ Action limits. The observed readings are plotted in the control chart and whenever the readings cross the warning limit, a retest needs to be taken to confirm the deviation. Once the repeated result also falls outside warning limits, then we need to take action for correction.

Special/Advanced Statistical Techniques

Chi Square test is used for comparison of end breaks in any Spinning machinery /Post spinning machinery like Ring doubling and TFO etc. Chi Square value for one degree of freedom is 3.84 as per Chi square table. Hence when we compare End breaks in a Ring frame against a standard or an expected performance value of another best Ring frame, then the,

Chi square value is
= Square of (O-E)/E

If the value is greater than 3.84, we can conclude that the Observed End breaks in the machine are significantly higher than the expected performance value.

F Test

F Test is normally used only when there is a need to compare the significance of the level of improvement from the control process. As an illustrative example:

TABLE 1.20

Example for "F" Test for Ring frame Count CV

Parameters	Ring frame 1-Control	Ring frame 1 –Trial
Count	60.0	60.0
CV % of count	1.60	1.35
No of leas tested	40	40
Standard deviation	0.96(SD1)	0.81(SD2)

F = (Square of SD1)/(Square of SD2)
= (0.96)/(0.81)
= 1.40

As per the statistics F table, for (n-1) degrees of freedom, F = 1.53

As the estimated value of F is less than Std "F" value of 1.53, we conclude that there is no significant improvement due to the trial process.

(In the same example if the CV of the trial process is achieved as 1.25, then estimated F value becomes 1.64, which is higher than standard "F" value and we can conclude that the trial process resulted in significant improvement).

Process Capability Estimate Cp

Process capability analysis is a set of tools used to find out how well a given process meets a set of specification limits. In other words, it measures how well a process performs in relation to the specification limits. The main objective of the process management is to prevent defects and to minimize the variability in the manufacturing process. To reduce the variability we need to establish the relationship between process variable and product results. Even though the product meets the specification limits and the process is in a state of control, the action may be initiated to quantify and further reduce the process variability to gain marvelous advantages. The statistical techniques can be used to quantify the variability. Process capability analysis helps in quantifying the process variability and assists manufacturing by eliminating or greatly reducing the variability.

$$\text{Process Capability Ratio Cp} = (\text{USL} - \text{LSL})/6\sigma$$

Process capability is inherent variation of the product turned out by the process. Process capability provides the quantified prediction of the process adequacy and refers to the "uniformity of the process." It is a measurement with respect to the precision of the Manufacturing process.

If the Cp is 1.0, the production process is just meeting the specified requirements.

If the value exceeds 1.33, the process is very much under control. (towards Six sigma level)

C_{pk} stands for process capability index and refers to the capability of a particular process for achieving output within certain specifications. In manufacturing, it describes the ability of a manufacturer to produce a product that meets the customer's expectations, within a tolerance zone. If C_{pk} is more than 1, the system has the potential to perform as well as meeting the requirements.

The equation for Cpk is [minimum(mean – LSL)/3σ, (USL – mean)/ 3σ], Cpk is used when the mean is not centered in the population.

If Cpk is greater than 1.0, then the process meets the specification limits and is capable too. If the ratio is less than 1.0, then the process has variation or the process is not stable.

Cp and Cpk are necessarily to be used by the mills during lot inspection and it helps the QA department to predict and control the manufacturing process effectively.

1.9 Quality Control Equipments in a Spinning Mill

In a Spinning mills, Quality control laboratory should have sufficient instruments for testing the Incoming lots by having testing instruments for fiber, testing instruments to evaluate in process materials like Sliver, roving, yarn in cop form, yarn in cone form and also testing equipments specifically for final yarn in packages.

1.9.1 Fiber Testing Equipments

Spinning mills need the following equipments for fiber testing so that the vendors can be evaluated and also to enable mixing issue by QA department.

Moisture content tester

By using moisture content tester mills check the moisture content of the incoming bales. The cotton fiber in the bales should have a moisture content between 7–8 % only as the standard moisture content of cotton is 7.5 % only. If the moisture content is higher, then the cotton will lose its moisture in the process/will offer resistance to opening /will result in higher imperfections in final yarn. Moreover it is a commercial issue too since we are paying for excess moisture content in the bales. Hence by random sampling every lot is inspected for Moisture content and reported to cotton procurement section upon the arrival of every lot. Digital moisture meters are available in the market with a needle/probe which is penetrated into the bales. Each bale must be tested at three positions-top, middle and bottom of the bales. Recommended sample size per lot is five % of the total number of bales per lot. The average of the total readings will be considered as the average moisture content of the lot. However if there are stray readings which may be very high like 9.0+ in some lots. In such cases we need to test more number of bales by sampling 10% of the bales per lot to arrive at our conclusion.

Trash Analyser

Trash in the cotton plays a major role in deciding the performance of the product in Blow room and Carding, end breaks in Ring frame, fabric appearance of the final product. Moreover it has a decisive role on deciding the Yarn realization of the product. Hence mill must check trash content in all the lots received as it has an impact on profitability of the company. Apart from the trash content measurement for the incoming lots trash measurement in in process material upto comber stage is also essential for the assessment of cleaning efficiency of Blow room cleaners/ beaters/openers, Cards and Combers. Mills use "Trash Analayser" for assessment of trash content in the cotton. With the help of modern Trash analyser we can assess the "microdust content" also apart from seed coat fragments, seeds and leafy particles.

HVI-High Volume Instrument

High volume instruments are used for the assessment of the Fiber length, Uniformity ratio, Short fiber content, Fiber strength, Fiber fineness and colour grade etc. The name is assigned as "High volume" mainly because the speed of testing in this instrument is very fast and the mills can test more than 200 plus samples in a shift practically. USDA standards of cotton classification is being followed for evaluation of fiber in this instrument. There are several manufacturers for this HVI or its equivalent instruments worldwide. This instrument has transformed manual cotton classification into instrumental classification swiftly.

With high volume instruments we can assess the colour grade which are characterized as "Reflectance and Yellowness."Both these properties are very useful in deciding the mixing issue for elimination of shade variation in the final product.

1.9.2 In-Process Materials Testing Equipments

Quality conscious Spinning mills who wish to excel in their performance need to have essential testing equipments as below for the estimation of quality characteristics of in process materials:

- AFIS-Advance Fiber Instrument System
- Evenness testers
- Lea strength testers
- Single yarn strength testers
- Twist tester
- Hairiness tester

AFIS-Advance Fiber Instrument System

Raw material accounts for 50–65% of the selling price of the Single yarn in 100% cotton depending on the count spun. Hence processing of cotton in Spinning mills need utmost care in handling the fibers at every stage of process from blow room to Speedframe. The processing machinery can damage fiber length and can create fiber neps etc. Moreover beaters in Blow room and lickerin rollers in a Card can lead to fiber rupture, if we don't set the process parameters properly.

AFIS PRO 2 tester is an instrument manufactured and marketed by Uster technologies. This instrument assesses the following characteristics of the fiber:

- Neps, Seed coat neps
- Fiber length-individual fiber length
- Fiber maturity
- Short fiber content by number and by weight
- Trash level

This instrument works on a laser beam principle to test the "individual fiber length of the sample presented to it" and the results are based on a minimum of 3000 fibers. Hence the results from this instrument is highly reliable. As it measures the individual fiber length of the sample the results on the length is reported as:

- 1% length- 1% of the fibers in the sample has a length above this length.
- 5% length- 5% of the fibers in the sample has a length above this length

1% length as tested from AFIS PRO is almost equal to the "Effective length" as tested by conventional Baer sorter method.

5% length is used by the mills for setting of drafting zones in Pre comber Draw frames, Lap formers, Combers and finisher Draw frames.

Short fiber content as measured by AFIS PRO is used for:

- Vendor evaluation for raw material procurement
- Setting of the process parameters like speeds and beater setting
- Deciding the noil level to be extracted in Combers
- Evaluation of Cards and Comber performance.

AFIS PRO instrument assesses the neps per gram present in Raw material as well as in "in-process materials like opened tufts, slivers in process, roving etc." Based on the changes in neps before and after any particular process changes in parameters need to be done to improve further.

AFIS PRO instrument helps to assess the "Maturity ratio" and "IFC-Immatured fiber content" of the cotton. These are essential paremeters of the raw material i.e. in Cotton as less maturity leads to "white neps" in the processed fabric. Moreover immature cotton fibers can lead to fiber damages and strength loss during in process.

Hence AFIS PRO is an essential instrument for "Process Management in a Spinning mill."

Sliver/Yarn Evenness Testers

In the Spinning mills, evenness assessment of the inprocess material is more important as it decides the "final yarn count and its mass variation." Evenness testers from its first introduction in the year 1948 has been tremendously developed to meet the dynamic needs of the spinners. First generation testers were able to detect "Evenness, thin(–50%), thick(+50%) Neps(+200%) only. As on today Evenness testers are able to estimate the following parameters by passing the materials through the Capacitive sensors at a set speed (different for Sliver,Roving and Yarn):

i. Higher sensitive Yarn imperfections(–30%, –40%, +35%, +140%)
ii. Spectrogram analysis
iii. Yarn Hairiness levels and its CV
iv. Mass variation CV in yarns
v. Sliver Evenness and Mass CV

As Evenenss testers help to test the variation level in mass/evenness of the material i.e. Sliver, Roving, Ring yarn in cop form, Yarn from packages like cones/cheeses, this tester offers a scope to improve the process for fine tuning Draft zone settings and also to find out the evaluation of Maintenance practices like Buffing, cleaning etc. Yarn can be tested at a speed of 800 m/min with auto feed arrangement. Testers have the facility of sucking the yarn being tested. Periodical variation caused by erratic eccentric movement of bottom rollers/top rollers can be assessed by this tester and few Instrument vendors including Uster also helps the mills to have "visual simulation of our yarn either in knitted or Woven fabric" too. This module helps the mills to assess the impact of periodcal variation and imperfections on fabric appearance to fine tune their process in a better manner. In Compact Spinning Ring frames, where Suction is used to compact the fleece, any variation between the suction level between the spindles will also lead to variation in yarn imperfections and hairiness. Hence evenness tester is an inevitable testing equipment in a Spinning mill.

Yarn Strength Testers

With respect to Yarn strength, testers are available for assessing the "bulk" strength of the yarn known as "lea strength" and also "individual yarn strength" which is universally known as "Single yarn strength."

Lea Strength Testers

Yarn from either the cops or cones are prepared in leas(Bunch of 80 threads of 1.5 yard each in a circular form prepared with the help of Wrap reel). Every lea is individually placed between upper and lower jaws of the Lea tester. With the help of a motor the bottom jaw will pull the leas downwards. At the point of breakage of leas, the strength will be recorded in the dial attached to the Lea tester. The dial is calibrated in "pounds" and hence the "Lea strength of yarn is expressed in pounds" only. In yarn trade, CSP-Count Strength Product is a well promoted Yarn quality characteristic which is used for Commercial/trading purpose. CSP is estimated by multiplying the Actual count of the leas tested in this instrument multiplied by the "lea strength in pounds." The modern Lea testers are having software to report the following data after completing all the samples:

1. Count (with the micro balance attached)
2. CV of Count
3. Lea strength of yarn
4. Strength CV

Single Yarn Strength Testers

Single yarn strength testers are used to test the individual strength and elongation characteristics of the yarn.

Spinning mills,as on today know that both quality and performance are essential if they need to meet the continually-rising demands of their customers in the Knitting/Weaving/Yarn dyeing. Apart from the appearance and less variations in imperfections Yarns must have the functionality to satisfy the end use running performance requirements at customer's end. Testers equivalent to Uster Tensorapid, helps the mills to estimate the Single yarn strength and elongation characteristics precise along with its variation levels. The report from these kind of testers cover the following Yarn tensile parameters:

- Single yarn Strength in grams/CV%
- RKm and its CV%

- Elongation and its CV
- Work of rupture

The priority for any yarn is the ability to withstand downstream processes – without causing stoppages or affecting production efficiency. Whatever be the fabric application, modern machines develops the greatest stress and strain forces on the yarn. Hence it is essential for the spinners to produce yarns with suitable strength and elongation properties for the fabric-making process, as well as for the ultimate end-use. Minimum strength and elongation properties are essential to prevent a yarn from breaking or being damaged in downstream operations, as well as avoiding blemishes on end-products in weaving. Hence accurate tensile-strength values are important, particularly for warp yarns, which are placed under tremendous stress.

Uster Tensojet 5 can operate at speeds of 400 m/min and it helps the mills to test the yarn for its tensile characteristics at very high speed levels to simulate the High speed weft insertion, High speed warping and Higher warp tension in Weaving. Spinning mills can assess the weak places in the yarn and fine tune the process to eliminate weak places in the yarn(or at least reduce it partially). This tester helps the mills to predict the yarn"s performance in Warping/Weft insertion in high speed looms. Tensojet instrument estimates and reports on the following tensile characteristics:

I. Single yarn Strength in grams/RKm in grams per tex/its CV%
II. Elongation and its CV%
III. Weak places(Elongation, Strength and total)
IV. Scatter plot of distribution of strength and elongation values which depicts "tensile health" of the yarn

1.9.3 Final Yarn Testing Equipments

During final inspection of the Cones/lot, necessarily samples will be randomly tested for Count and its CV, Lea strength CV, Single yarn tensile characteristics using the instruments discussed above. However for the assessment of fabric appearance, we need to evaluate the yarn for its Classimat faults which occurs randomly and has decisive impact. Hence Classimat testers have been introduced in the Spinning mills to evaluate the random faults remaining in the yarn after clearing by the clearers in the Autoconers. However, it is not an essential equipment in the spinning mills today as Modern clearers from the leading vendors like Uster and Loepfe provide voluminous data on random faults and periodic faults including contamination levels.

1.10 Final Lot Inspection of Products

Final lot inspection is a routine event for a normal mill. But in the mills who wish to delight the customers and excel in performance, we should definitely take care of the following precautions:

I. Building the quality into the product as per the customer's specific requirements
II. Tracking the same through monitoring Incoming inspection of Raw materials, In process materials at critical stages and periodical Machine Wise Rogue spindle analysis in Ring Frame
III. Setting alarm limits for specific characteristics based on Customer requirements in Autoconers. For example if the customer is a knitter for high end bleached white fabric, we need to assure 100% elimination of colour fibers in knitted and bleached fabric or minimizing the incidence to acceptable limits to the customer. Even though we take enough care in Raw material selection Blow room contamination segregation we need to adopt proper settings in Autoconers for Foreign fibers selectively since it is the final stage of the process.
IV. Conducting simulation studies for meeting the requirements of end use performance characteristics. For example, if the customer is a direct warper, then necessarily we should conduct a warping simulation study at our end in Autoconer to find out the "breaks per million meters" to evaluate the "yarn and package quality."
V. Inspecting all the cones physically under adequate illumination conditions over a black base table, so that no cone defect misses our eyes.
VI. Over and above, we should draw a statistically significant number of samples from the lot and inspect the cones for all the characteristics we committed to the customer without fail.
VII. Certain expectations from the customers are implied one like low CV of cone weights, absence of mildew on the surface, fumigated pallets etc which must be taken care of during final inspection.

Lot inspection reports during final inspection are important documents for the organization and they are supposed to be kept safely for future reference for any kind of customer feedback/Complaints. A model "Final inspection Lot report" is enclosed in the Annexure VIII

1.11 Key Performance Parameters (Quality) in each department

The QA department must establish key performance parameters for each department, which could vary between customers too and it has to be tracked by the department with warning and action limits. Following table covers a few key performance parameters which can be taken only

TABLE 1.21

Key Performance Parameters (Quality)

Department	Material	KPP	Remarks
Blow room	Delivered chute material	Trash in tuft	
		Neps per gram	
		Drop in 2.5% fiber length	
Cards	Sliver	Sliver trash	
		5 m CV%	
Comber and its preparatory	Lapformer	Initial and final layer Hank CV	
	Comber Sliver	Neps per gram	
		Sliver U%	
		SFC(N)	
Drawframes	Sliver	Sliver Evenness	
		1 m CV%	
Speed Frames	Roving	3 m CV%	
		Neps /gram	
		Roving evenness	
Ring frames	Yarn	Count CV	
		Lea strength CV	
		Single yarn strength CV	
		P(0.1) value in Tensojet	
		Ratio of P0.1 to Mean strength	
		Hairiness CV	
		CV of imperfections	
Autoconers	Cones	Specific classimat faults	
		Cone weight CV	
Autoconers	Cones	Rewinding breaks per million meters	
	Knitted fabric-(white bleached and dyed)	• Colour fiber per kg	
		• Knitted fabric-Dyed	
		• Colour fiber per k	
		• White pp/kg	

for the guidelines since Customer requirements and expected inprocess quality requirements are dynamic in nature.

This list of KPP is only a sample covering more critical and important based on Customer's requirements but few customers requirements will be more specific on a single parameter itself like "CV of Hairiness" alone depending on their end use. Those customers expect the same specification needs to be complied by us–apart from meeting all other standard specifications. In those circumstances mills must be very cautious to focus our activities in all areas which have an impact on Hairiness and its CV like fiber rupture in Blow room, Humidity conditions in Departments, Traveller changing frequency, Compact zone conditions etc.

CS1. Case study on "Higher level of End breaks in 40s Combed Compact Warp yarn"

In a mill, where 40s Combed Compact yarn is produced for the weaving market, we suddenly experienced the breakage level of 7 breaks per 100 spindle hours instead of less than 4 breaks per 100 spindle hours consistently. Hence mixing quality and all the preparatory in process parameters were analysed for deviations. Upon analysis we found that there is a steady trend of increase in average trash content in the mixing laid from 2.3 to 3.2 in a period of seven days. The following deviations were identified in the in process results:

- Blow room delivery trash has been deteriorated from standard level of 1.2 to 1.5
- Card Sliver trash content has been deteriorated from the level of 0.02 to 0.08

As the mills use a Suction compacting system with perforated aprons, Sliver trash content in Card Sliver plays a major role in deciding the end breaks. Hence it is immediately attended at Blow room with seed trap adjustments and waste fine tuning at Uniclean. When the trash content level in Card input was reduced to below 1.2 by making these changes, mills were able to reach the target level of less than 4 breaks per 100 spindle hours.

CS2. Case study on "Higher Unevenness in Finisher Draw frame sliver"

In one mill, 30s Combed Compact yarn is a regular product. In draw frames, suddenly there was an increase in Sliver unevenness from 2.1 to 2.4 and the mills followed a Bottom roller setting of 38/42 as per their mills process parameters. The machine and process parameters were investigated along with Mixing characteristics. We identified the changes in Raw material quality and in-process result as below:

- SFC(N) in mixing had a change from the higher level of 29% to 25%
- Average 2.5% Span length of the fiber at mixing stage had gone up by 0.4 mm
- In Comber sliver 5% length had a change from 38 mm to 39 mm

Mills followed a principle of Bottom roller setting in Draw frames as below:

Front bottom roller setting = 5% AFIS length in combed sliver
Back bottom roller setting = Front zone setting +4 mm

Hence the mills bottom roller setting became inadequate for the increased length of 1 mm in combed sliver which caused the hike in unevenness of Draw frame sliver. Hence the bottom roller setting was corrected to 39/43 mm settings instead of 38/42 and trials were conducted subsequently to confirm improvement in Sliver unevenness. QA department confirmed the resultant Sliver unevenness is 2.05 against the target level of 2.1.

CS3. Case study on "Increase of Clearer cuts in Autoconer in 60s Combed Compact weaving yarn"

A spinning mill planned to manufacture 60s Combed compact weaving yarn for export and the initial clearer cuts were established at 85 as below:

Neps - 7
Short thick places - 40
Long thin places - 3
Long thick places - 3
Count channel cuts - 4
CCP&CCM Cuts - 8
FD Cuts - 20
Total cuts - 85

Suddenly the mills experienced the higher level of clearer cuts in all the drums and it was inconsistent too with a wide range fluctuating between 140–160. Upon scrutinization of clearer cuts we identified that the increase in cuts is contributed by increase in Short thick places mainly. Hence, mixing quality and in process results are analysed in depth. We identified the following major deviations:

- Moisture content in two lots,which were laid in the mixing, had a moisture content of 9.5%. Average moisture content in the mixing laid has gone up by 0.6%.
- Nep level in cards were found to be in the upper limit in the majority of the cards.

Hence the cotton bales from these lots were opened and dried in the cotton godown itself. After drying for two days the moisture content is measured and found to be 7.4%. These opened materials are re-baled and given to mixing. Bulk level studies are followed up for clearer cuts in Auto coner with these dried up lots and we found short and long thick places are brought to normal conditions as earlier.

CS4. Case study on "Increase in foreign fiber(FD) cuts in Autoconers"

In a mill, which is spinning 80s Combed compact weaving yarn for High end Airjet looms, mills maintained a level of FD cuts (Foreign fibers-Dark) at 25. Suddenly the QA department observed the FD cuts showed a trend of increase from 25 to 32. Upon investigation, the findings are:

- No significant difference in Contamination index in the mixing used
- No changes in the operators used for contamination cleaning

However mills identified one inconsistency in their contamination clearer ejection level in one of the blow room lines. The ejection level was set at 2000 ejections per hour and the mills used to collect 0.4% ejected waste per day.

During this particular period of analysis it was identified that the ejection per hour was reduced to 1400 per hour and the ejected waste was also reduced to 0.3% instead of 0.4%. The contamination clearer unit was thoroughly checked and found failure of solenoid valves for ejection. It was immediately corrected and ejections per hour was normalized to 2000 per hour. This fresh production was followed up upto Autoconer stage and the FD cuts were normalized to the standard level of 25.

CS5. Case study on "Increase of Yarn Count CV in 40s Combed Hosiery yarn during final lot inspection."

In a mill where 40s Combed Hosiery yarn is spun in 20000 spindles we identified the count CV was at a higher level of 1.4% when the lot was inspected. Fortunately the mill follows a very good channelization system and the assignment was easier to identify the origin of defect. Upon shop floor investigation from packing to Ring frames we identified a set of Ring frames catered by one Speed frame has resulted in a higher CV of 1.7%. Hence the 3 m CV % of Roving was tested for the particular Speed frame immediately and we found the result as 1.6% against the mills target of 1.3%. A thorough audit was conducted on Drafting system wherein we identified:

- Deviation in top arm pressure between spindles
- Non alignment of top rollers with bottom rollers in few spindles

The Speed Frame drafting system was attended by the maintenance team and the fresh bobbins were tested for 3 m CV%. It was improved to 0.9% and hence the yarn count CV from these set of particular Ring frames were brought to the standard level of below 1.2%

Annexure I: Quality Control Department Organogram for a Spinning Mill with 50000 Spindles

		No of personnel
Head-QA		1
QA Officer		1
Incoming Raw Material investigator (HVI, Trash separator)-1	Inprocess Material investigator (AFIS)-1	2
Process Auditor (Cleaning efficiency, NRE, CN%, Waste level)-0.5	Inprocess Material Investigator (Evenness and Imperfection)-2	2.5
Process Auditor (End breaks, Cop rejection, Clearer cuts, Rogue spindles, Rewinding studies)-0.5	Routine Hank and Count inspection Assistants (shift wise)-3	3.5
Final Lot inspection investigator-1	Special Studies and Customer follow up Assistant-general shift-1	2

Note: For the position of investigators, auditors and assistants we need relievers @ one per every three persons. Hence we need 4 more persons as relievers. Apart from Head QA and one QA officer, we finally need 14 trained persons to manage the QA department.

Annexure II: Model Quality Control Schedule for Mills Striving for Excellence in Manufacturing

No	Department	Activity	Sample size	Frequency
1	Cotton-Incoming inspection	Fiber quality-HVI	5 bales per lot	For every incoming lot
		Fiber quality-AFIS	5 bales per lot	For every incoming lot
		Trash content	Sample of 100 grams evenly from 5 bales per lot	For every five lots of a vendor
		Moisture content	5 bales per lot	For every incoming lot
2	Cards	Hank of sliver and 5 meter CV	6 balls per card with 5 m for each ball	All cards once in a fortnight

(*Continued*)

No	Department	Activity	Sample size	Frequency
		Evenness testing	125 m/card	Once in a month
3	Precomber Drawframe	Hank of sliver	6 balls per delivery	Weekly once
		Evenness testing	125 m/Delivery	Once in a fortnight and also after Buffing of cots
4	Lapformer	Lap Hank CV –initial and final	1M*4 Samples for initial and final	Once in a fortnight
5	Comber	Sliver hank and 5 m CV	6 balls per Combers with 5 m for each ball	Once in a fortnight
		Evenness testing	125 m/Comber	Once in a fortnight and also after Buffing of cots
6	Drawframes	Hank of sliver and 1 m CV	6 balls of 5 m each	Once in a Day
		Evenness checks	125 m/Delivery	Once in a week and also after buffing of cots
7	Speedframe	Stretch and 15 m CV and Hank checking	Initial layer 20 bobbins/final layer -20 bobbins (15 m per bobbins)	Once in a fortnight
		Evenness	125 m/Bobbin for 4 bobbins	Once in a fortnight and after buffing
8	Spinning Frame	Count and CSP	20 cops per count of a group of five RF`s	Once in a week
		Evenness and Imperfections	8 cops /RF @400 m/min for 2.5 min	Once in a fortnight and after buffing
		Twist checking	10 cops per side of RF from separate spindle tape position	Once during count change and tape change
		Single yarn strength	10 cops per side of RF from separate spindle tape position	Once during count change /mix change
		Snarl testing	5 Cops per RF	Only for warp counts and HT counts/ during count change

(*Continued*)

No	Department	Activity	Sample size	Frequency
Special Studies				
1	Blow room	Cleaning efficency	Beater wise	Once in 3 months and after mixing change
		Nep generation	Beater wise	Once in 2 months and after mixing change
2	Cards	Nep removal efficiency	Every card	Once in a month and after wire change
		Zone wise waste study	One hour of card production	Once in 2 months and also after mixing change
3	Comber	Comber noil checks	All the eight heads	Once in a fortnight
		Nep removal efficiency	All the eight heads	Once in a month
		Combing efficiency	All the eight heads	Once in 2 months
4	Drawframe	Nep level	All the deliveries	Once in a month
		Improvement in 5% AFIS length	All the deliveries	Once in a month
		A% Checks	Both plus 1 and Minus 1	Once in a fortnight
5	Speedframe	3 m CV	20 separate bobbins /15 m each	Once in 2 months
6	Ring frames	Count CV –RF wise	20 cops per RF	Once in 2 months covering all RF channelized for a Speedframe
		Online classimat studies		Everyday recording from Autoconer clarers
		Rogue spindle studies-RF wise	Every RF –All spindles	Once in 2 months
7	Autoconers	Splice quality	10 drums per day	Once in 3 months to cover all the drums
		Drumwise rewinding studies	Only for warp counts	All drums to be covered once in 3 months
		Analysis of clearer cuts	All Autoconers	Shiftwise from reports
		Cone weight CV	All drums	Once in a month
8	Yarn conditioning	Gain in moisture content	All counts	Once in 2 months
9	Packing section	Bag/Carton weight (Nett and tare weight)	Count wise-20 bags/cartons	Once in 2 months

Annexure III: Estimation of Contamination Index in Raw Material

Annexure III									
	Estimation of Contamination Index							Date	../../....
Cotton Vendor						**Variety**			
Name of the ginner						**Station**			
Name of the presser, if different						M.L.NO			
Broker:	**ABC COTTON COMPANY**					PR-NOS			
Bale No.		6		30		**Average Contamination Index(C.I)**			
Bale weight in Kgs		170		170		170		**Norms**	**Historical data**
CONTAMINATIONS									
	Weightage	Nos	Wt in gm	Nos	Wt in gm	Nos.	Wt in gm		
GRADE A (Weightage: 40%)	**40**								
1. White PP string									
2. White PP fabric									
3. Alkathene bits									
Average		**x**	**y**	**a**	**b**				
C.I (Grade A)		**= (X*40+Y*40)/200**		**= (a*40+b*40)/200**		**Average**		**0.40**	
GRADE B (Weightage: 30%)	**30**								
1. Coloured PP string									
2. Coloured PP fabric									
3. Hair									
4. Metal Piece									
Sub Total		**c**	**d**	**g**	**h**				
C.I (Grade B)		**= (c*30+d*30)/200**		**= (g*30+h*30)/200**		**Average**		**0.20**	

(Continued)

Bale weight in Kgs		170	170	170	Norms	Historical data
CONTAMINATIONS						
GRADE C (Weightage: 20%)	**20**					
1. Coloured yarn						
2. Coloured fabric						
3. Coloured cotton						
4. Coir						
5. Feather						
6. Oily Cotton						
Sub Total		**e f**	**i j**			
C.I (Grade C)		**= (e*20+f*20)/200**	**= (i*20+j*20)/200**	**Average**	**0.20**	
Sub Total (A+B+C)						
C.I (Grade A+B+C)					**0.20**	
GRADE D (Weightage: 10%)	**10**					
1. Supari Cover						
2. Black cotton						
3. Stone						
4. Jute Cloth						
5. Jute twine						
6. Wooden piece						
7. Stick						
8. Leather piece						
9. Others						
Sub Total		**k l**	**m n**			
C.I (Grade D)		**= (k*10+l*10)/200**	**= (m*10+n*10)/200**	**Average**	**3.60**	
					3.80	

(*Continued*)

Over all C.I (Grade A+B+C+D)						
		Excellent	Very Good	Good	Average	Poor
Industry Norms	Contamination index (A+B)	0	0.5	1	1.3	1.6
	Contamination index (A+B+C)	0	0.8	1.4	1.8	2.5

Note: Normally Contamination Index (A+B) is used for Imported cottons only. Contamination index (A+B+C) is used for all types of Indian cotton varieties. Grade D is less harmful and normally used only for Indian cottons and it helps to negotiate on prices based on level of such contamination.

Annexure IV A: Evolution of Clearers

Yarn produced from Ring frames has many defects which will harm aesthetic appeal of the fabric and also affect its performance during end use like Knitting, Warping, weft insertion or direct yarn dyeing. Hence it is essential to clear such defects in winding process so that the customers get the product with the least defects. Initially, mills used mechanical slub catchers (with steel or ceramic blades which were set based on yarn diameter) to catch and clear the defects – only thick places and slubs which ahs more than one eighth diameter of yarn.

In Spet 1958, Thomas Holt & Co exhibited the first ever electronic yarn clearer `Wilson Yarn Clearer`in the International Trade Exhibition at Manchester, UK developed by Mr.Peter Wilson, Glascow.

From 1960 onwards Uster has started manufacturing `Electronic yarn clearers` and its development are chronologically given below:

1. Uster Spectomatic
2. Uster D Type clearer
3. Uster Automatic C/W
4. Uster Polymatic clearer
5. Uster Payer Clearer
6. Uster Quantum clearers(1999)

As on today Uster Quantum has released its UQ4 –fourth version of Uster Quantum clearers with 'DUO' clearing facility, launched in 2021.

Mechanical slub catchers cleared only thick places and most disturbing faults like accumulated fluff with less than 50% clearing efficiency. But today's electronic yarn clearers has the following facilities to ensure the products meet the customer's expectation on end use performance as well as aesthetic appeal of the fabrics.

Features available in the latest yarn clearers comprise:

1. Coarse/Fine count with a defined range
2. Short thick places as per smart clearing curve
3. Elimination of periodic cluster faults
4. Elimination of long thick/long thin faults
5. More hairy yarn
6. Colour contamination
7. Transparent and opaque foreign fiber contamination
8. Splice defects

9. Shade variation assessment and elimination
10. Alarm limits for defects
11. Online imperfection measurements
12. Online Classimat faults measurement
13. Storage of data
14. CV of imperfections

UQ4 clearers have been provided with 'DUO CLEARING' facility with capacitance as well as optical clearing sensors. Due to this facility it can be set for improving the visual yarn appearance too.

Ref: Comparative analysis of Uster Electronic Yarn clearers of the last ten years by Rahim Umer Uster News Bulletin.

Annexure IV B: Classimat Faults

Yarn defects

With the increased pressure on higher yarn quality requirements and improvements in evenness levels a better method is in need to identify what is tolerable and what are disturbing defects-or "Outliers". Uster Classimat provides a new visual representation of the yarn, which is called "YARN BODY".

YARN BODY is the illustration of the normal yarn with its expected natural mass variations and its tolerable and frequent yarn faults. The faults that do not belong to the YARN BODY are considered `non frequent and most probably "disturbing while processing the yarn".

YARN BODY changes with,

- Raw material
- Spinning process

Visual representation of yarn body in the tester has,

- N,S,L,T events
- Green dots –remaining events
- Yarn Body variation in shaded area
- Yarn Body-Dark green shaded area
- NSLT outliers

NSLT is given in 45 classes in Classimat testers out of which NSL is covered by 30 classes. Classimat faults are broadly classified into three major categories :

- Short thick faults (Sum of A,B,C and D)
- Long thick faults(Sum of E,F,G)
- Long thin faults (Sum of H1 and I1)

No	Classification of defect	Origin of Defect
1	Short thick faults	• Higher trash in mixing • Less cleaning efficiency in process • Immatured cotton • Poor tuft opening efficiency • Improper settings in critical zones in Cards • Absence of clearer devices/Rollers in drafting systems • Improper selection of spacers and condensors • Damaged wire points in Cards and Combers • Poor quality of brushes in Combers
2	Long Thick Places	• Mixing containing unopened rovings • Improper bottom roller settings in drafting zones • Aged or incorrect aprons in Speedframes • Improper piecing in Drawframes and Speedframes • Bottom rollers` "felting" in Speedframes and Ring frames • Accumulation in flyers • Mixing with wide variation in length • Higher roving twist multiplier • Lower level of Back top roller load • Hose leakages in Pneumatic drafting systems
3	Long Thin Places	• Web falling in Cards • Strip damage in Doffers(Cards)/Cylinder wires • Higher creel draft in Drawframes and Speedframes • Lower roving twist multiplier • Roving stretch • Sliver splitting in Speedframe creel • Incorrect fit of Roving and Ringframe tubes • Higher tension level in Autoconers

Conclusion: Mills must identify the exact origin of defects as listed above and ensure suitable steps are taken to reduce the level of defects or exploring the possibility of eliminating the defects(in case it is more severe) through proper Clearer settings in the Autoconers.

Annexure V: Package Defects

Introduction

As a manufacturing organization the entire team of any spinning mills sweat hard to produce yarns meeting customer`s requirement in all respects. However even if the yarn is produced with all necessary Cutomer's stated quality requirements, finally it has to reach the customer and perform satisfactorily at his immediate end se process, say, knitting, warping, dyeing or as direct weft. Autoconers and packing section are vulnerable departments to meet the Customer's requirement in meeting the end use process requirements.

Common failures in Autoconers and packing which could lead to certain kind of Package defects are listed below alongwith remedial measures. Based on the experience in the industry, package defects are divided into two categories viz "Visible and Invisible."

No	Package defect	Causes	Remedial measures
Visible Package Defects(VPD)			
1	**Soft cone**	**Less tension applied in Autoconers** **Vibration in Cone holder/drums** **Air leakage in Cradle parts**	Tension assembly checking and correction Educating and training the tenters on the impact of Vibrating cone holders and correction Systematic air leakage monitoring management
2	**Hard cones**	**Higher tension in yarn tensioner** **Higher cradle pressure**	Optimisation of yarn tension relevant to the count and product Establish standards for type of products and its enduse
3	**Ribbon wound cones (Fig 1)**	**Failure of Anti ribboning mechanism** **Damaged grooves in drums**	Periodic checking of Anti ribboning mechanisms by shop floor Finding out the root causes for damages and eliminate recurrence as this is very difficult to identify the damages in drums

(Continued)

No	Package defect	Causes	Remedial measures
4	**Stitches in Cones (Nose and base) (Fig 2)**	**Incorrect settings of cone holder to drums** **Inconsistency in the paper cone specifications, especially the length** **Yarn path disturbance** **Defective yarn guide/ traverse in drumless drives in Autoconers**	Systematic resetting procedures in Autoconers for all critical settings Establish strict `Incoming packing material inspection` for checking against our specifications Audit the machine`s critical parts like Tension assempbly, yarn guides, traverse guides etc and take corrective measures
5	**Cut ends in cones** (Fig 1.3)	**Drum damages** **Damaged surfaces in Material handling trollies** **Excessive tension** **Weak places in yarn** **Weak splices**	• Periodic drum inspection • Optimization of process parameters in Autoconers • RF wise analysis of rogue spindles • Drum wise analysis of higher clearer cuts • Splice appearance and strength analysis drum wise periodically • Material handling trolleys must be inspected fro surface smoothness once in a fortnight and transporting employees must be educated to look into it.
6	**Collapsed cones**	**Very less tension to the cones** **Tight packing** **Improper material handling** **Shrinkage of paper cones during yarn conditioning** **Poor strength of Paper cones**	• Optimize process parameters for yarn count and customer requirements • Train and educate material handlers • Paper cone specifications to be restudied and corrected
7	**Nose bulging in cones**(Fig 1.4)	**Cone holder to drum setting disturbance** **Lower level of pressure at Nose** **Improper traverse in drumless drives**	• Periodic drum setting and cradles/cne holders • Air leakage checks in Autoconers • Traverse driving elements drive checks periodically
8	**Nose flowering in cones**(Fig 1.5)	**Higher yarn tension at nose**	• Optimized tension during winding

(Continued)

No	Package defect	Causes	Remedial measures
		Poor paper cone quality	• Incoming inspection for all specifications of Paper cone
9	**Bell shaped cones** (Fig 1.6)	**Higher yarn tension**	• Higher yarn tension • Failure of tensioning unit
10	**Multiple tail ends/ Absence of tail end**(Fig 1.7)	**Paper cone specifications**	• Tail end groove and other specifications need to be maintained • Surface finish of Paper cone must not be smooth
		Autoconer tenter work practices	• Tenters must inspect tail end`s presence randomly in the cones doffed • In critical end uses tail end should be pasted with a tiny sticker
11	**Mildew in Yarn surface**(Fig 1.8)	**Moist conditions of storageMoisture content more at the time of packing**	• Godown storage conditions must be maintained warmer to maintain the Moisture content of yarn. • Moisture content in the yarn at the time of packing should not exceed 7.4% • QA Department should check the physical appearance of cones once during two months during rainy seasons and also in Humid environment.
Invisible Package Defects(IPD)			
12	**Entanglement in cones**(Fig 1.9)	**Higher level of repeaters during Autoconer winding**	• Abnormal clearer cuts need to be checked drumwise and attended to. • Waste clogging in suction arm needs to be eliminated • Suction level in suction arm needs to be optimized • Suction arm gripper needs to be checked for quality
		Higher level of Hairiness CV	• Hairiness CV needs to be maintained below 5% in all types of yarn • In compact yarn CV of hairiness between spindles must be checked and corrected.
13	**Waste bits inside the cones** (Fig 1.10)	**Improper cleaning of yarn path**	• Autoconer cleaning with proper cleaning solution needs to be done

(*Continued*)

No	Package defect	Causes	Remedial measures
		Throwing of waste from OHTC/Diffusers from Humidification plants **High level of repeaters leading to layer disturbance/yarn accumulation** **Untidy upper arm /less suction level in upper arm** **Malfunctioning of Balcon** **Failure of Antikink device** **Improper pre clearer settings**	• OHTC Cleaning and Diffuser cleaning at scheduled intervals must be followed up • Drums with high level of repeater cycles need to be analysed daily and attended • Rogue drum must be identified by machine reports in display or through available reporting systems like Mill master, Cone expert or Visual Manager • Preclearer settings must be optimized and checked at periodic intervals • Antikink device must be checked by shopfloor foreman • Balcon centering and its movement must be checked periodically
14	**Nep latching during rewinding** (Fig 1.11)	**Higher level of incidence of fibrous/seed coat neps in yarn**	• Process parameters must be optimized for controlling Seed coat neps in Sliver at Cards • A2 and B2 classimat faults need to be minimized to the extent possible to eliminate the incidence of Nep latching.
15	**PP Contamination**	**Higher level of PP contamination in raw material**	• Mills should establish CI* Estimation and start incentivizing the vendors • Based on CI, Manpower for manual contamination cleaning must be done • In the case of Contamination clearer units in Blow room, settings must be optimized for effective removal of PP contamination. • In Autoconers, PP clearing should be fine tuned. • QA must have a regular contamination assessment through analysis on grey/bleached and dark dyed knitted fabric.

*Note**: CI- Contamination Index (Pl refer Annexure III).

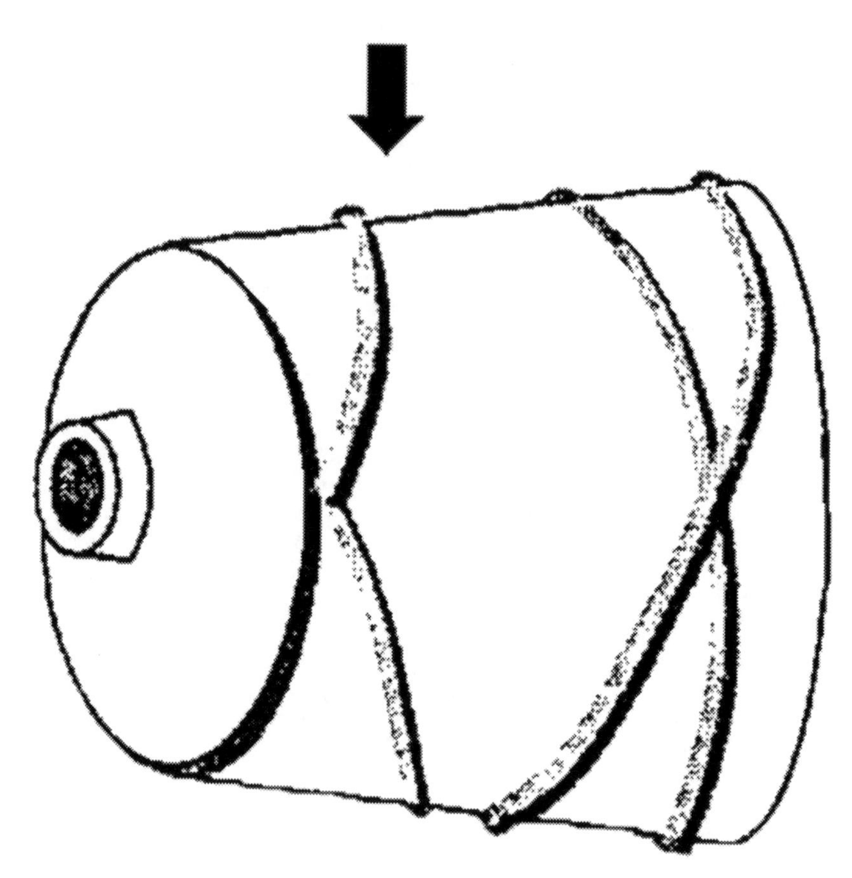

FIGURE 1.1
Ribbon Wound Cones.

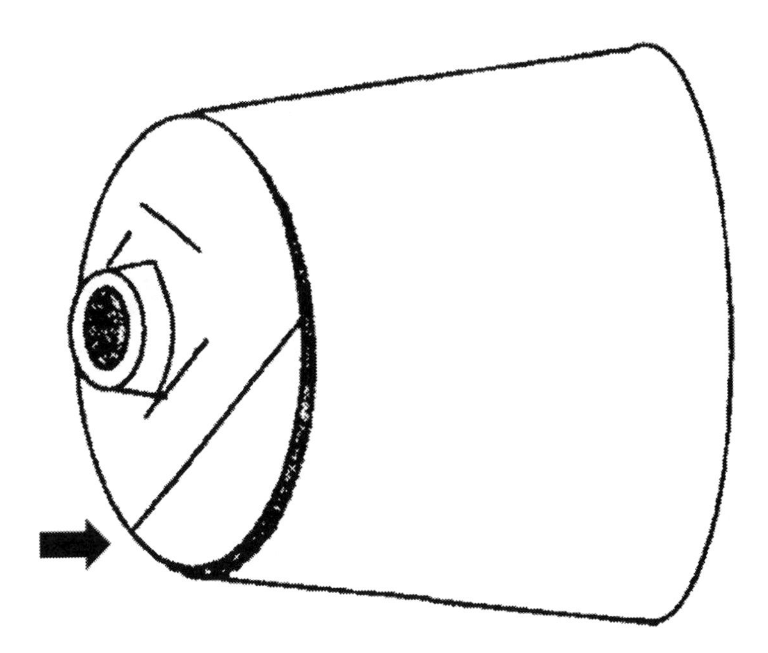

FIGURE 1.2
Stitches in Cone (Nose).

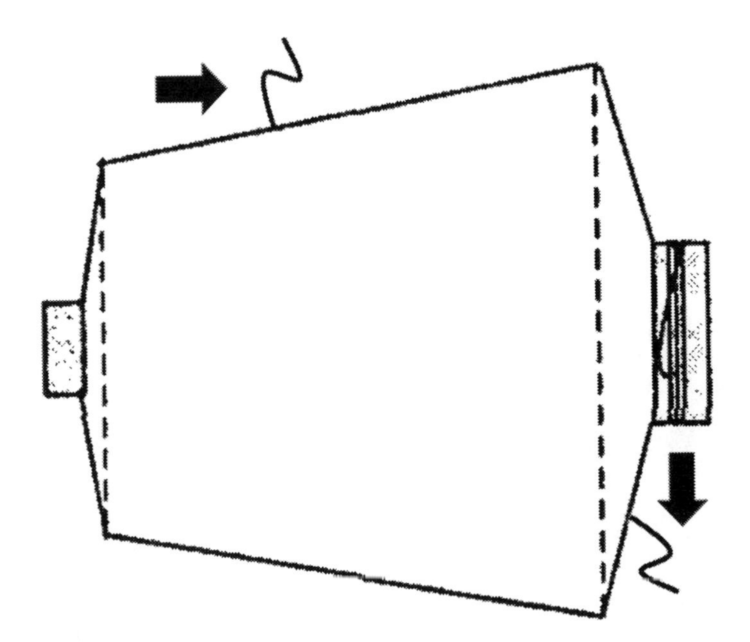

FIGURE 1.3
Cut Ends in Cone.

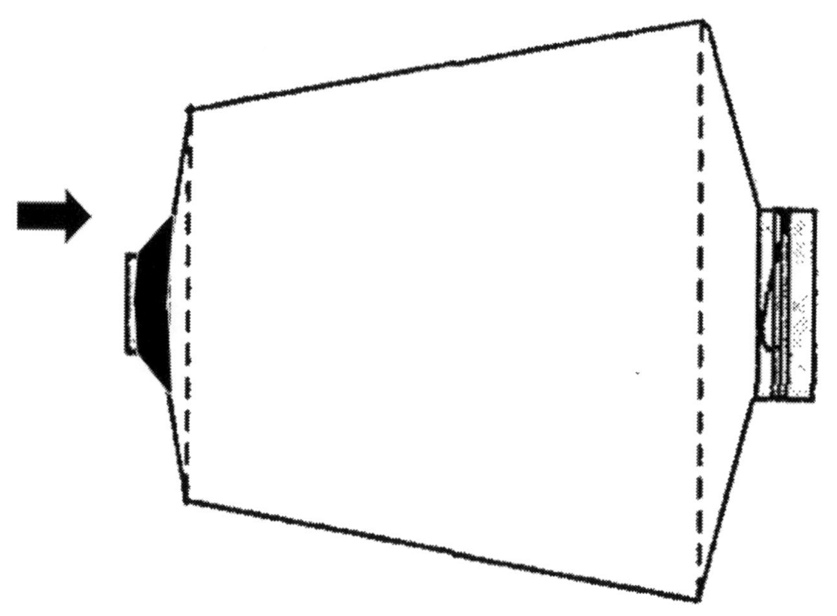

FIGURE 1.4
Nose Bulging.

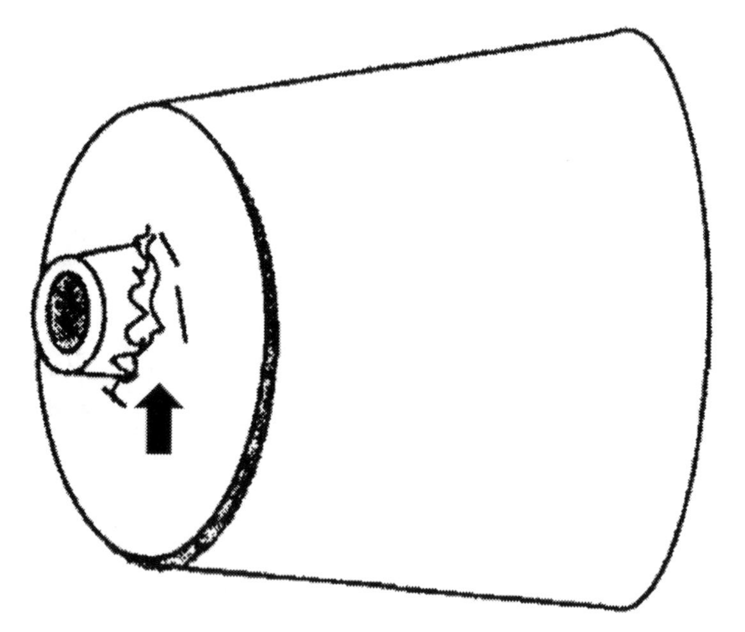

FIGURE 1.5
Nose Flowering.

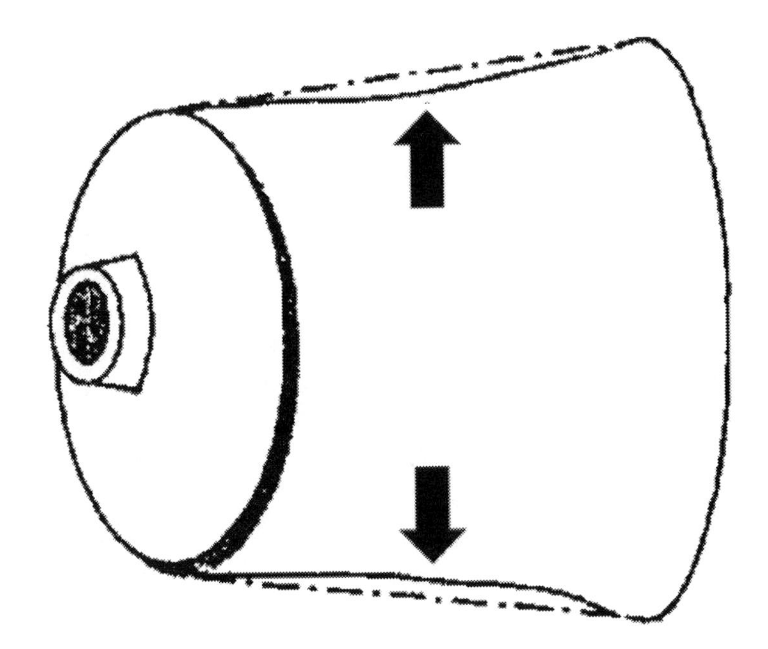

FIGURE 1.6
Bell-Shaped Cones.

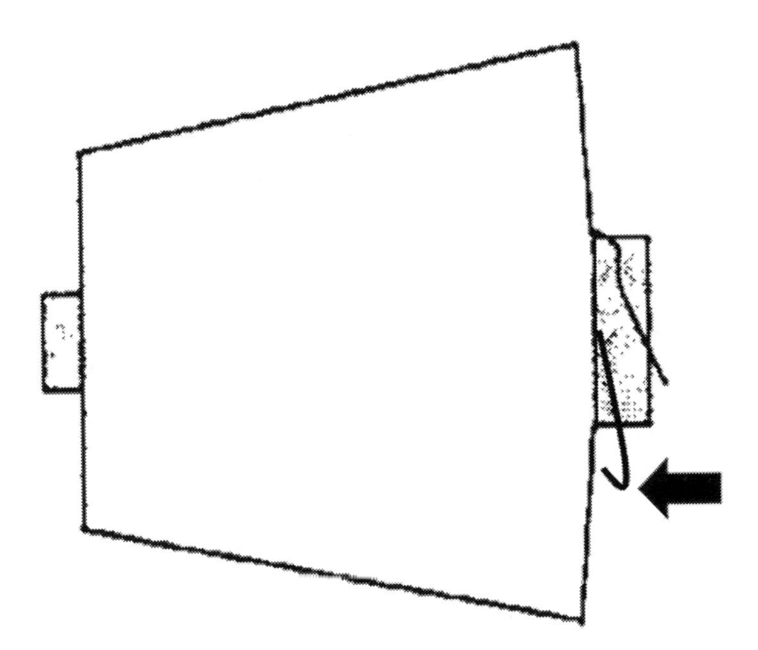

FIGURE 1.7
Multiple Tail Ends.

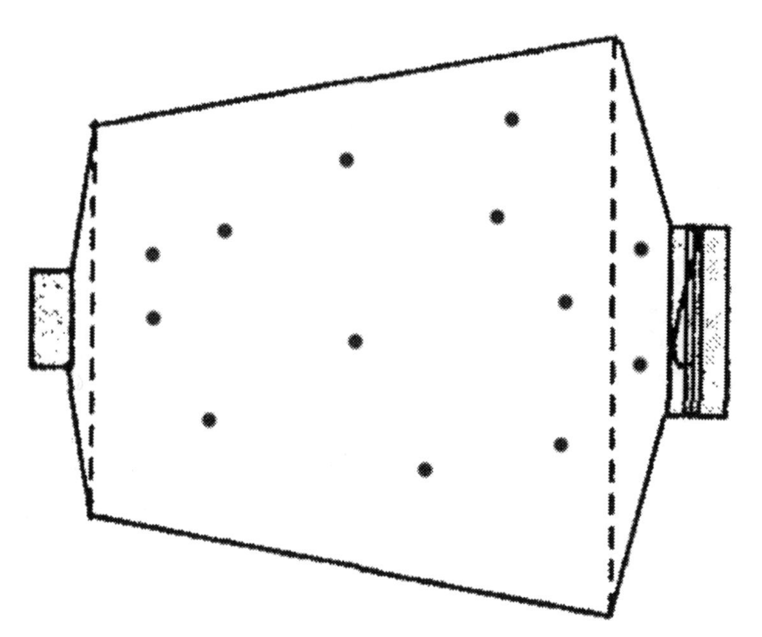

FIGURE 1.8
Mildew in Cones.

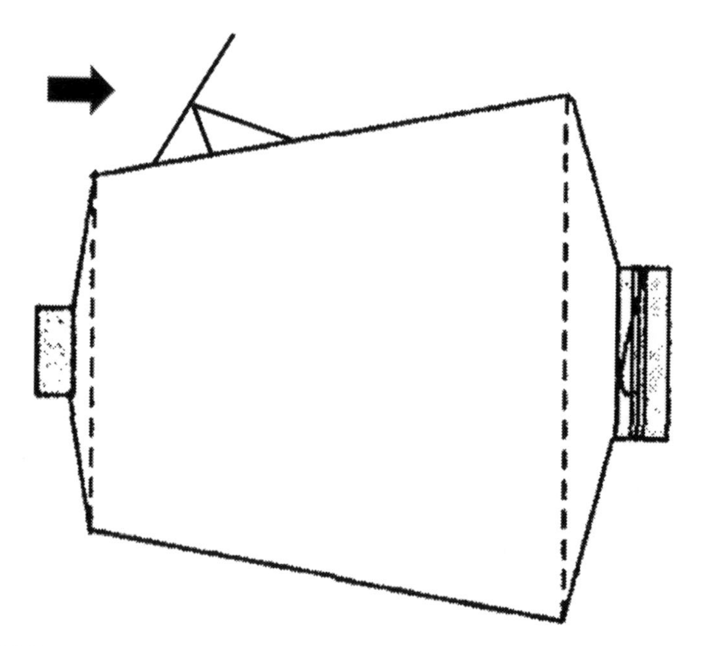

FIGURE 1.9
Entanglement in Cones.

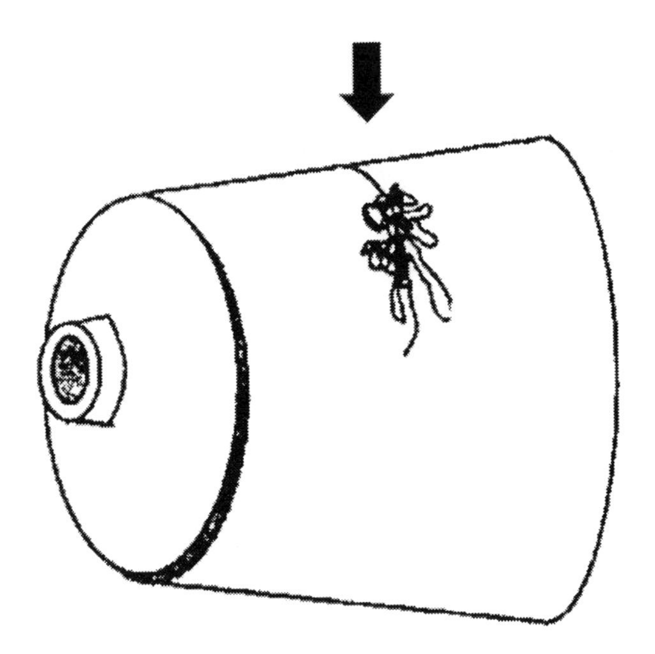

FIGURE 1.10
Waste Bunches in the Cones.

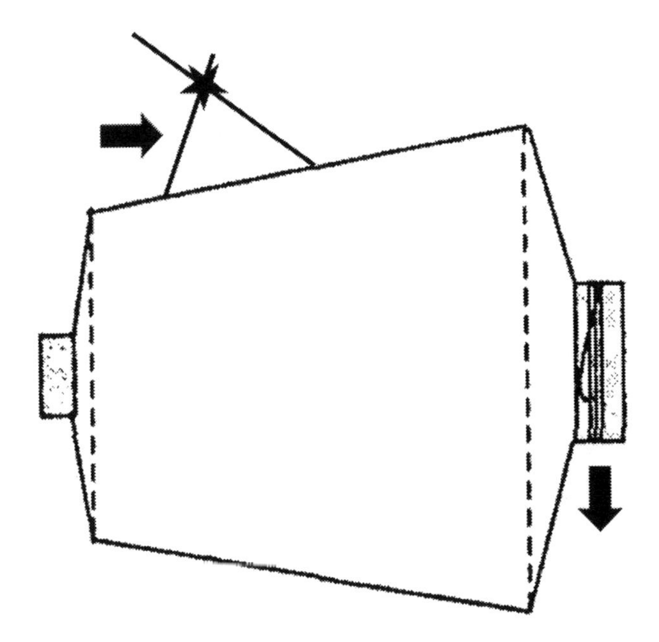

FIGURE 1.11
Nep Latching in Cones.

Annexure VI: Process Audit format

Department/Machines	Mills Std	Actual Parameters
Blow room		
Blendomat		
Mixing laydown order		
Depth of penetration		
Use of water spray		
Line 1 and 2 -Common equipments		
Uniclean settings		
Line 1		
Beater settings		
Contamination clearers-Ejections per hour trend		
Line 2		
Beater settings		
Contamination clearers-Ejections per hour trend		
Cards		
Production rate in kg/hr		
Cylinder speed		
Flat speed		
Lickerin speed		
Tension draft		
Can content in length		
Auto leveller settings		
Pre comber drawframe		
No of ends		
Trumpet size		
Identification system		
Channelization		
Lap formers		
Lap length		
Identification system		
Channelization		
Combers		
Speed in nips per min		
Production rate in Kg /hr		
Sliver hank		
Trumpet size		
Top combs settings		
Identification system		
Channelization		

Department/Machines	Mills Std	Actual Parameters
Draw frames		
No of ends up		
Auto leveller settings		
1 M CV%		
Top rollers interchange procedure		
Trumpet Size		
Scanning roller specifications		
Identification system		
Channelisation		
Speed frames		
Spacers used		
Spindle speed		
Bobbin weight		
Condensors-Front		
Channelisation		
Ring frames		
Spacers used		
Cop content		
Speed pattern		
Twist per inch(display)		
Traverse status		
Identification system		
Channelisation		
Autoconers		
Clearer settings		
N		
S		
L		
T		
Cp		
CCp		
CCm		
Foreign fibers(FD)		
Foreign fibers(white PP)		
Cluster settings		
Cone weight-Actual by 20 cones weighment		
Cone length settings		
Identification system		
Channelisation		
Drum speed -Actual		
YCP		
Count of yarn being conditioned		
Programme settings		

Annexure VII: Quality Audit format

Department/Machines	Good	Average	Poor	Remarks
Blow room				
Blendomat				
Magnet cleanliness				
Condition of water/oil spray				
Quality of droppings in seed trap				
Line 1 and 2 -Common equipments				
Metal detector droppings				
Heavy particle separator droppings				
Common cleaner droppings at AWES				
Line 1				
Beater droppings quality at AWES				
Unimix/Multimix-filling pattern				
Contamination clearers-Ejected cotton quantity				
Line 2				
Beater droppings quality at AWES				
Unimix/Multimix-filling pattern				
Contamination clearers-Ejected cotton quantity				
Cards				
5 M sliver CV				
Quality of droppings at AWES-Lickerin				
Quality of droppings at AWES-Flat strips				
Sliver appearance				
Doffed can -Spring quality				
Pre comber drawframe				
Doffed can -Spring quality				
Sliver appearance				
Draft zone cleanliness				
Lap formers				
Lap appearance				
Lap-last layer quality from Combers				
Draft zone cleanliness				
Combers				
Web appearance in individual heads				
5 M Sliver CV				
Quality of Noil at AWES				
Doffed can -Spring quality				
Sliver appearance				
Draft zone cleanliness				

Department/Machines	Good	Average	Poor	Remarks
Draw frames				
Sliver appearance				
1 M Sliver CV				
Doffed can -Spring quality				
Sliver appearance				
Draft zone cleanliness				
Speed frames				
Creel sliver alignment				
Function of top clearer unit				
Function of bottom clearer unit				
Draft zone cleanliness				
Roving flow of movement				
Bobbin density-by feel				
Bobbin appearance				
Ring frames				
Restarting breaks				
Idle spindles				
End breaks level				
Traveller loading				
Tapes running above wharve				
Spindles with lower twist(if ISM is avbl)				
Rogue spindles(if ISM is available)				
Autoconers				
Drums below std level of performance				
Total clearer cuts				
Tension breaks				
Mis splicing%				
Upper yarn sensor failure				
Kg/Hour				
Foreign fiber cuts				
White PP cuts				
Auto doffer Mis-Doffing				
Cone weight CV%				
Package defects in each Autoconer-From packing section				
Shade variation in cones-Type of product wise				

Annexure VIII: Final Inspection Lot Report Model Format

Count
Customer name

No of lots despatched
Total Quantity despatched

Date of Inspection

S.No	Yarn quality characteristics	Unit	Mill Stds	Customer`s specs	Current results-lot no X	Previous lot results				Average of Previous four lots
						Lot (X-1)	Lot (X-2)	Lot (X-3)	Lot (X-4)	
1	Actual Count									
2	Count CV	%								
3	CSP									
4	Lea strength CV	%								
5	Twist									
6	Twist CV	%								
7	U	%								
8	Thin places(−50%)									
9	Thick places(+50%)									
10	Neps(+200)									
11	Total Imperfections/km									

Final Lot Inspection Report

Count Customer name					No of lots despatched Total Quantity despatched				Date of Inspection	
S.No	Yarn quality characteristics	Unit	Mill Stds	Customer`s specs	Current results- lot no X	Previous lot results				Average of Previous four lots
						Lot (X-1)	Lot (X-2)	Lot (X-3)	Lot (X-4)	
12	Total Higher sensitive Imperfections/km									
13	S3 Value of Hairiness									
14	Hairiness									
15	CV of Hairiness	%								
16	RKm									
17	CV of RKm	%								
18	P(0.1) of RKm									
19	Ratio of P(0.1) to Mean RKm									
20	Elongation	%								

Final Lot Inspection Report

Count Customer name					No of lots despatched Total Quantity despatched					Date of Inspection
S.No	**Yarn quality characteristics**	**Unit**	**Mill Stds**	**Customer`s specs**	**Current results-lot no X**	**Previous lot results**				**Average of Previous four lots**
						Lot (X-1)	**Lot (X-2)**	**Lot (X-3)**	**Lot (X-4)**	
21	CV of Elongation	%								
22	Total Weak places									
23	Objectionable faults									
24	Long this places									
25	Long thick places									
26	Total classimat faults									

Final Lot Inspection Report

Comments

InvestigatorQA HeadHead of Manufacturing

Count **Customer name**					**No of lots despatched** **Total Quantity despatched**			**Date of Inspection**		
S.No	**Yarn quality characteristics**	**Unit**	**Mill Stds**	**Customer`s specs**	**Current results- lot no X**	**Previous lot results**				**Average of Previous four lots**
						Lot (X-1)	**Lot (X-2)**	**Lot (X-3)**	**Lot (X-4)**	
	End use performance simulation studies									
1	Foreign fiber per kg									
2	Colur fiber in Bleached fabric									
3	White polypropylene in dyed fabric									
4	Breaks per million meter in Rewinding studies									
	Process capability Cpk									

2

Maintenance Management in Spinning mills

LEARNING GOALS

- **Types of Maintenance**
- **Lubrication**
- **5S**
- **FMEA**
- **TPM**
- **Jidoka**
- **Machinery Maintenance Audits**
- **ZDQC**
- **PDM techniques and tools**
- **SPDM**
- **SPMF**

Introduction

Good Machinery maintenance not only contributes for high productivity but also can be considered as a vital contributor to the product quality. Neglect of Maintenance and the timely replacement of parts/critical components will prove several times expensive compared with the costs associated with a proper maintenance programme. In some cases, a very small wear can contribute to a disproportionate loss, if not replaced in time. For example an oil seal for a splicer in Autoconer will result in severe warping breaks and rejection of a lot itself which is deadly.

Machineries are expected to be maintained to get the best output in terms of Productivity and Quality economically. Moreover these machineries must contribute with the maximum utilization possible/designed

DOI: 10.1201/9781003257677-2

by the manufacturers. Machinery manufacturers give recommended procedures for Maintenance of their machinery but shop floor maintenance crew are expected to modify to suit the organizations` requirements. In this executive handbook, we cover the important elements of Maintenance with a focus on Spinners who deal with Premium products & who wish to contribute for excellence in manufacturing.

2.1 Types of Maintenance

There are several schools of thought on types of Maintenance. Simple bifurcation for a better understanding as per the Industry practice at global level is elaborately dealt with.

2.2 Breakdown Maintenance-First Generation Maintenance

The maintenance activities carried out as and when the machine breaks down on an emergency basis. This is not at all a good practice and mills try to eliminate such types of maintenance by analysing the root causes of such breakdowns in detail to eliminate in total. Role of Machinery manufacturers needs to be understood by giving timely feedback to the Machinery manufacturers for permanent correction. However this was the first generation Maintenance and still being used by few Small scale spinning units and Ginneries in India.

2.3 Planned Maintenance-Second Generation Maintenance

Activities of maintenance that are carried out to prevent any defects in the product or breakdowns, to retain the life of spares and the machine as a whole, restore the life of critical components etc. Planned maintenance can be further categorized as below:

- Routine Maintenance
- Preventive maintenance

2.3.1 Routine Maintenance

Routine maintenance covers all activities which are carried out in a routine manner with a periodicity for retaining the life of machinery and its spares as per machinery manufacturer's recommendation. However mills depending on their count pattern, location and type of employees modify these schedules to meet their own conditions without forgetting the purpose behind each activity.

TABLE 2.1

Critical Routine maintenance activities -Department wise

No	Department	Activities	Frequency (days)
1	Blow room	General cleaning	Daily
		Full cleaning	15
		Resetting	90
2	Cards	General cleaning	Once in 2 days
		Petrol stripping	Once in 3 months
		Flat tops cleaning	Once in a week
		Full setting	Once in 6 months
		Grinding	As per QA studies
3	Lap formers	General Cleaning	2
		Full cleaning	21
		Buffing of top rollers(once in three full cleaning)	63
4	Combers	General cleaning	2
		Full cleaning	21
		Buffing of Draw box top rollers(once in two full cleaning)	42
5	Drawframes	Draft zone cleaning	2
		Full cleaning	15
		Buffing of top rollers-All lines	15
		Scanning rollers/Autolevellers setting checks and correction	30
6	Speed Frames	Draft zone cleaning	3
		Full cleaning	30
		Flyer cleaning	90
		Top roller buffing	60
7	Ring frames	Draft zone cleaning	2
		Full cleaning	21
		Top roller cots buffing	63
		Mango tubes cleaning(for compact RF)	5

2.3.2 Lubrication

Lubrication in Spinning mills has seen a sea change in the last three decades. From the level of conventional lubrication systems with manual intervention today`s machinery parts are designed with centralized lubrication system, Bushes and bearings are filled with long life greases (even up to 4-5 years), Gear boxes only with oil topping requirements etc. With the advent of Sustainable greases, technology offers environmentally friendly lubricants which ensure a pollution free environment.

As the technology is fast changing one with new products to suit High speed machinery along with energy conservation techniques, mills are advised to have an audit of their current lubrication systems and proposed system of lubrication/lubricants with the help of Original equipment Manufacturers or Kluber/Kluber equivalent lubricant suppliers. During the audit mills need to evaluate the quality of the equipment used for lubrication like Oil cans, sprayers for petrol, Sprayers for Autoconer lubrication, Centralized lubricant tanks and hoses/tubes etc. More importantly the personnel engaged for Lubrication must have a thorough awareness of the products and lubricants and machinery parts lubricated- needs to be trained by experts from Lubricant vendors.

Expected output from an expert Audit on Mills Lubrication:

- Adequacy of the Lubrication schedule and suggested modification
- Premature wear of driving elements including shafts/bearing/housing and gearboxes
- An inspection report on the lubrication tools like grease guns/grease pumps/centralized lubrication system etc and measures to correct the same
- A performance report on the Lubrication crew
- Storage system of Lubricants and the inventory level/control
- Suggested advanced long life lubricants to rationalize the Lubricants and also control the cost of Lubrication

Based on the expert`s audit report mills must take suitable corrective action on utilizing the advanced technology of lubrication by streamlining the schedules and changing the type of lubricants. We also should utilize the lubrication experts to train our Lubrication crew also for effective implementation.

2.3.3 Replacement programme

Mills should follow the guidelines as specified by the machinery Manufacturers for replacement of critical parts/spares.However it must be borne in mind mills need to review the same periodically as the life of the spares depend on the `conditions in which we operate the machines (environment conditions/speeds/utilization levels/type of fibers used etc).

General guidelines for replacement of critical items are provided in the table below:

TABLE 2.2

Replacement programme for Critical spares

No	Department	Name of the part	Units	Life span (Range)
1	Blow room	Pinned beater	Years	4–7
		Saw toothed beaters	Years	1–2
		Beater shaft and bearings	Years	10–12
		Feed rollers to beaters	Years	8–10
2	Cards	Cylinder wire	tonnes	1000–1200
		Flat tops	tonnes	600–800
		Lickerin wire	tonnes	200–250
		Pre cleaning segment	tonnes	200–250
		Post cleaning segment	tonnes	600–800
3	Combers	Unicomb	Years	6–8
		Top comb	Years	2–4
		Brushes	Years	2–3
		Detaching rollers with cots	Years	3–4
		Drawbox top roller cots	Years	1.5–2
		Drawbox Top rollers	Years	4
4	Drawframes	Top roller cots	Months	6–9
		Top rollers with bushes	Years	3–4
		Cogged belts replacements	Months	6–9
5	Speed Frames	Top roller cots	Years	1.5–2
		Top roller shells	Years	6–8
		Clearer cloth	Years	2
		Top and Bottom Aprons	Years	1.5–2
		False twister	Years	1.5–2
6	Ring frames	Top roller cots	Years	1.5–2
		Top and Bottom Aprons	Years	1.5–2
		Top roller shells	Years	6–8
		Spindle tapes	Years	1.5–2
		Bobbin Holders/Lappets	Years	4–6
		Pneumafil suction tubes	Years	5–7
		Rings	Years	5–8
		Spindles	Years	8–10
7	Compact System	Mango tube Aprons	Months	8–10
		Inserts-Coarse counts	Years	1.0
		Inserts-Medium counts	Years	1–1.5
		Inserts-fine counts	Years	1.5–2
		Intermediate gears	Years	2–3

2.4 Preventive Maintenance

Preventive maintenance (PM) covers all activities done to the machinery in order to keep them running at its maximum capability level in terms of Productivity and Quality and also to prevent any unexpected major breakdowns leading to financial loss to the organization.

Benefits of Preventive Maintenance

- Higher levels of reliability and life span of Machinery/Parts
- Higher utilization of Machinery with least breakdowns
- Less rejects due to Quality deviation in the products from these machinery
- Ensure safe work environment to all the employees

TABLE 2.3

Critical Preventive Maintenance Activities

No	Department	Activity	Frequency
1	Blow room	All dust cages and dampers-cleaning and polishing	6 m
		All grid bars checking for free movements	3 m
		All beaters-settings & condition	3 m
		Suction level checking	15 days
2	Cards	AWES Suction level	15 days
		Wire point inspection	3 m
		Autoleveller inspection	3 m
		All the revolving rollers/beaters condition/ Auto Doffing	6 m
		Undercasings	6 m
		Bare flats and flat chains	1 year
3	Lap formers	All pneumatic parts checking including Clutches	6 m
		Top roller cots buffing	3 m
4	Combers	Pneumatic lines and parts checking	3 m
		Nipper condition checks	6 m
		Critical parts like Ledges/Top comb brackets/ Brushes	During every full cleaning
		Auto Doffing mechanism checks	3 m
		Detaching roller cots Buffing	Not advisable
		Drawbox cots buffing	2 m
5	Drawframes	Top roller cots buffing	15 days
		Top roller pressure checking	6 m
		All kinds of sensors and Scanning roller functions	3 m

(*Continued*)

TABLE 2.3 (continued)

Critical Preventive Maintenance Activities

No	Department	Activity	Frequency
		Bottom fluted roller bearings inspection	1 year
6	Speed Frames	Top roller cots buffing	3 m
		Bottom fluted roller bearings inspection	1 year
		Top Arm pressure checking	1 year
7	Ring frames	Top roller cots buffing	30 days to 45 days
		Bottom fluted roller Eccentricity and bearing checks	1 year
		Pneumafil suction level checks	2 m
		Spindle centering	1 year
8	Compact System	Mango tube suction level checks	1 m

2.5 Systems to enhance Maintenance of Assets/Machinery in Spinning mills

Machinery Maintenance Audits are one of the primarily established system in Spinning mills which help the mills to Audit and improve their

- Maintenance schedules
- Replacement frequency
- Type and Schedule of Lubrication
- Benchmarking our maintenance programme
- Control on Inventory

By critical examination and audit of our Machinery conditions with a structured checklist, investigation of our current Maintenance schedules, replacement programme, Maintenance equipment like Grinding accessories, Buffing equipment and accessories and tools & gauges to evaluate its adequacy. Machinery maintenance audit also covers investigation of Quality trends of critical in process performance parameters like NRE%, Cleaning efficiency, CV of U%, Range of Imperfections, Range of clearer cuts etc so as to pinpoint the failures of related maintenance activities which spoiled the particular quality.

Model check list for effective Machinery Maintenance Audit–Offline and Online are enclosed in Annexure IX A and IX B respectively.

Apart from Machinery Maintenance Audit, there are established systems developed by leading companies like Ford and Toyota on systems to improve Manufacturing efficiency through shop floor improvements and analytical approach in engineering and Automobile industries which are very useful for Spinning mills too. Three such systems are:

- 5S System for Shopfloor excellence
- FMEA –Failure mode effect Analysis
- TPM-Total productive Maintenance

2.5.1 5S System for shop-floor excellence

5S is a technique for shop floor housekeeping developed and promoted by TOYOTA. 5S is a system for organizing spaces, so that work can be performed efficiently, effectively, and safely. This system focuses on putting everything where it belongs and keeping the workplace clean, which makes it easier for people to do their jobs without wasting time or risking injury.

5S -Explanation

The term 5S comes from five Japanese words which are translated as:

- **Seiri** - Sort&dispose
- **Seiton** - Set in order/Organize
- **Seiso** - Cleanliness(cleaning with a purpose of inspection)
- **Seiketsu** - Standardize (establish Standard Operating Procedures)
- **Shitsuke** - Sustain (by establishing Auditing systems/Motivational schemes)

5S System helps the industry to improve shop floor housekeeping/ Service rooms and stores with an orderly arrangement/Offices are well maintained for easy retrieval of files/Tools and gauges in an accessible location for efficient management of time etc.

The important tools in 5S system are:

- Red tag area: Where excess or unwanted material in a particular department is kept for the inspection of other departments. After inspection by everyone concerned, these materials will be sorted out as `Materials to be retained and materials to be disposed`.
- Fixed point Photography: 5S coordinators will take photographs of a particular spot (where they plan to show improvements).

Photographs will be taken by standing in a fixed location/angle of observation. These photos will be displayed in the shop floor in a 5S Notice board and all employees will be requested to suggest for improving this workspot. All the ideas generated will be pooled and the best suggestions will be implemented. After implementation, FPP will be taken once again to assess the improvements. Again these photographs will be displayed and employees need to suggest further improvements if any. This system will be reviewed periodically for improving further and sustaining the same.

- Visual display system: 5S demands visual display systems for locating the spot where a material is stored/where an office is located etc. This system helps everyone to `value` the time which is precious.
- Visual Management System: Under Visual Management System, we expect grand improvements through visual and effective communication systems. The benefits of VMS are:
    - It helps problems/issues/improvements visible and transparent
    - Status is communicated effectively
    - Improvement in focus areas is made easier and timely actions by the responsible persons
    - Instills confidence among the teammates/cross functional team
- SOP-Standard Operating Procedures: SOP`s need to be formulated once we standardize our practices to be followed up based on 5S implementation at every stage. 5th `S` speaks about sustainability of the system in the organization for which we need to create SOP once the practices are implemented successfully. Few SOP`s in the Spinning mills where 5S is successfully implemented are:
    - Scrapping the unwanted material to Scrap yard in separate bins material wise
    - Keeping the `Materials to be disposed` in a Disposal yard with proper tagging/Maintaining registers
    - Handling the external visitors
    - Maintenance of wash rooms and inspection schedule
    - Inspection of Service rooms
    - Evaluation of 5S teams periodically

2.5.2 FMEA-Failure Mode Effect Analysis

A Failure Mode and Effect Analysis is an engineering technique used to define, identify, and eliminate known and potential failures, problems, errors and so on from the system, design, process and Services before they reach the customer.

During the Manufacturing operations we encounter Breakdown of Machinery/Repeated rejects in process/Customer complaints etc. For identifying the root causes and to solve it permanently we can utilize this FMEA tool. The essence of the FMEA technique has three components which help to define the priority of the failures:

- **Occurrence(O)** - Frequency of the failure
- **Severity(S)** - Seriousness of the failure(Effects/impacts)
- **Detection(D)** - Ability to detect the failure before reaching the customer

Based on the above three factors, RPN-Risk Priority Number is estimated by the multiplication of all these three. RPN has no standard value and it can only be used as ranking. Commonly we use 1 to 10 rating scale since it provides ease of interpretation, accuracy and precision. Detection and Rank is tabulated for FMEA analysis:

TABLE 2.4

Severity Guidelines -FMEA

Detection	Rank	Detection	Rank
No effect	1	Significant effect	6
Very slight effect	2	Major effect	7
Slight effect	3	Extreme effect	8
Minor effect	4	Serious effect	9
Moderate effect	5	Hazardous effect	10

FMEA process

We must follow a systematic approach to conduct FMEA effectively.

I. Team selection and Brainstorming: In any such FMEA, we should ensure concerned cross functional members are available in the team so that an effective brainstorming can be conducted within the team. One of the tools used for brainstorming is `fishbone diagram-cause and effect analysis`.

II. Functional block diagram/Process flow charts: In the case of System or Design FMEA we can use block diagrams and for Process/Service FMEA`s we can use Process flow charts. These charts vividly explains the inter-relations with cross functional teams, components, processes, assemblies etc.

III. Prioritize: Once the problem is identified, we need to prioritize `where to begin`.

IV. Data collection: Now the team needs to collect the data on failures and categorize them appropriately. The failures identified by the team are the `failure modes of FMEA`.

V. Analysis: Once the data is made available, the team must utilize the data for resolution using their knowledge and analytical skills.

VI. Results: Theme under FMEA is `results are data driven`. Based on the analysis of data results are derived. RPN can be identified based on the analysis of Occurrence, Severity and detection.

VII. Evaluation: Once the results are known, we need to evaluate the level of improvement.

VIII. Do it all over again: Irrespective of the level of achievement in step VII, we should try to do it all over again for further improvements as FMEA is a tool for `Continual Improvement`.

FMEA is more useful in Breakdown analysis and mills must use this technique effectively by analysing the breakdowns in critical machines as the first step. Both mechanical and electrical breakdowns can be analysed through FMEA.

2.5.3 TPM-Total Productive Maintenance

Total Productive Maintenance is one of the acclaimed systems for achieving Manufacturing excellence from Shop floor itself by involving everyone in the organization.

2.5.3.1 Origin of TPM

Total Productive Maintenance, **TPM**, originally had been started as a method of physical asset Management focused on maintaining and improving the Manufacturing machinery, in order to reduce the operating cost in an organization. The JIPM (***Japanese Institute of Plant Maintenance***), expanded it to include 8 pillars of TPM that required involvement from all areas of manufacturing in the concepts of lean

Manufacturing. TPM is designed to disseminate the responsibility for maintenance and machine performance, improving employee engagement and teamwork within management, engineering, maintenance, and operations.

TABLE 2.5

Eight pillars of TPM

Abbreviation	Expanded pillar	Focus area
AM	Autonomous Maintenance	Machine operatives diagnose the causes for the losses in efficiency and contribute for improvement
FI	Focused Improvement	Scientific approach to problem solving to eliminate the losses
PM	Plant Maintenance (Professional Maintenance)	Professional maintenance activities inside the plant by trained and competent personnel
QM	Quality Management	Scientific and Analytical approach for problem solving for defect analysis and ensure ZDQC
IC	Initial Control	Creative and innovative approach on equipment and its design concepts to eliminate losses and make it easier to make defect free production efficiently.
E&T	Education & Training	Training systems and support for continuous improvement of knowledge and Competence level of all employees in the organization
OTPM	Office and Administration TPM	Use of TPM tools to improve all the support functions to the core Manufacturing operations
SHE	Safety Health and Environment	Systems to improve/sustain the Safety of all the employees hygiene conditions and Environmental conditions conducive to Society

2.5.3.2 TPM implementation

Once the organization decides to implement TPM across the entire organization, it can arrange for an extensive TPM training programme with experts from TPM implemented organizations or through TPM Experts. This training programme must be tailored to suit top level executives/Cross functional management team/Shop floor management team/Shop floor employees and all other employees in support functions including office. This training programme must be utilized to 'assess the status quo' of our organization in different scales. A "Gap Analysis" (Annexure X) by TPM experts can help the organization to do a better assessment. Once this is completed, TPM organogram (Annexure XI) must be formed with individual responsibilities for each pillar. Based on the gap analysis,

Pillar chairman and its team members should brainstorm themselves in periodic meetings at shop floor(as far as possible) and bring out scope for improvement areas and action plans with a defined target date. Mills can conduct periodic reviews and submit reports to the top management once in 3 months to assess improvements. Management, at its own discretion can decide the intervention of an external expert assessment for further improvements, if there is a need.

Autonomous Maintenance(Jishu Hozen)

As the first step of implementation of Autonomous Maintenance we need to first declare a 'Manager Model Machine' in anyone department. Let us assume one of the Card in the Carding department is selected for the same. This machine will be first inspected for 'status quo' conditions in all respects by the responsible cross functional executives alongwith the Carding tenter and the maintenance crew. During the inspection process everyone will inspecte the machine for dirt deposits/oil leakages/air leakages/suction drop/worn parts/production efficiency/ Quality results etc. A brain storming session will be conducted at shop floor itself and the Carding tenter will be motivated to contribute more as he is the user of this machine for 8 hours in a day. Based on the findings of the team members, tagging of the parts where improvements need to be done will be carried out. An action plan will be drawn to attend the tagged spots in the machine. Once the actions are completed a review will be carried out by the team in the same manner to evaluate the effectiveness of the implemented actions and scope for further improvement will be discussed and finalized. This process will be continued until the selected Card reaches the expected level of performance as per our defined targets and the operator should be able to contribute for the maintenance of the machine.Once the necessary actions for autonomous maintenance of Card is finalized, SOP-Standard Operating Procedure will be drawn for JH of Cards. It is the responsibility of the JH team to horizontally implement in all the similar cards in the Carding department and review periodically. Similarly other machines also will be brought under TPM –JH network by the team. Autonomous Management helps the operator confident of maintaining the machine and operating at its maximum efficiency by his involvement.

Individual Improvement team needs to unearth the reasons for Utilization loss of the assets. It needs to analyse the underlying causes for Breakdowns/downtime due to any reasons including change overs or labour induced stoppages/Quality rejects/Idle spindles etc. The focus is to maximize the overall Machine utilization. Overall Equipment Efficiency(OEE) is to be focused and targeted for all machines in the department.KK team(…)must unearth the causes for the loss of OEE in all the cirtical machines on top priority and establish SOP for improving and sustaining the Overall Equipment Efficiency.

Planned Maintenance (PM) pillar team must work towards improvement of reliability of Assets(Machinery and equipments) through Design improvement/Condition monitoring techniques/Operator training for elimination of defects/Multi skill training to operator to forecast and avoid breakdowns. Basic tools of PM pillar is MTBR(Mean time between repair) and MTTR(Mean time to repair) and through this analysis PM team can create standard operating procedures for improving the reliability of the assets which help the mills to improve Productivity/ Quality/Utilization levels.

E&T (Education and Training) pillar needs to focus on Operator and employee training to 'prevent/eliminate any kind of deviation in performance of the machines which will lead to sub standard performance'. The technological experts on the machinery from OEM should be arranged to educate the employees in such kind of knowledge dissemination. E&T team also should use Customer samples (Complaints/ feedback/development) to educate and train the concerned department employees for defect prevention. This will help the organization to build a strong dedicated and involved team of employees in the long run. Employees will be motivated to do 'Kaizen' projects for enriching their knowledge and enhancing their contribution.

QM team will develop its team members for simulation of 100% inspection by as otherwise impossible in a spinning mills. It needs to create internal complaints/feedback system and resolve it like an external customer complaint so that external customer complaints are totally eliminated. Fail safe practices/Defect prevention/ZDQC/JIDOKA are the tools for this pillar which are explained elsewhere in this book too.By all these means `COQ-Cost Of Quality` must be reduced to the lowest possible level.

Initial Control pillar focuses on new product developments in the shortest span of time and also achieving standard operating conditions when the mills procure and commission new machinery/equipments. The responsible team will analyse the probable cause for delays based on the past experience (own or other equivalent spinners) and take all measures beforehand to prevent delays in development of new products/commissioning of new machinery or equipments.

OTPM (Office administration TPM) pillar focuses on the quality of support service to the manufacturing team by following standard operating procedures like JIT in procurement function/Lean Management in HR function etc.

SHE (Safety, Health and Environment) pillar team needs to concentrate on all activities concerned with the Health and Safety of all the employees, machines and the factory premises by focusing on the measures essential towards the same by reviewing with the concerned team members. Accidents of any kind will be analysed in depth and evaluating the levels of performance by indicators like `SIN` (Safety Index Number).

2.5.3.3 ZDQC-Zero Defect Quality Control

ZDQC is covered in detail under QM pillar as one of the prime activities in TPM organization and it covers all activities that cover the Quality control activities which helps the Maintenance and QA engineer work together to build the Quality into the product by prior assessment of critical in process quality characteristics.

Spinning mills manufacture products meeting Bench marked quality requirements or Customers specific requirements. In all these products, we need a process plan to achieve final product quality without any rejects. This plan covers all the in process checks that are necessary for the ZDQC system.

TABLE 2.6

In-process checks for ZDQC

No	Department	Activity	Impact on
1	Blow room	Nep generation level	• Yarn Imperfections • Classimat faults
2	Cards	Nep removal efficiency and Neps per gram	• Yarn Imperfections • Classimat faults
		SFC(N) analysis	Yarn strength
3	Combers	Web appearance	Yarn imperfections
		SFC (N) Analysis	Yarn strength
4	Drawframes	1 m CV%	Final yarn Count CV and Strength CV
		Sliver Unevenness	Final yarn Imperfections and Autoconer clearer cuts
5	Speed Frames	Stretch %	• Classimat faults • Yarn strength CV
		Breaks per 100 spl hours	• Lean cops in RF • Yarn strength CV
		3 m CV%	Count and Strength CV
6	Ring Frame	Rogue spindle analysis	Weak spots in yarn
		Evaluation of cots buffing	Yarn imperfections and weak spots
7	Autoconer	Clearer cuts analysis	• Classimat faults • Weak spots
		Splice quality analysis	End use performance improvement
		Rogue drum analysis	Package defects and End use performance

All these activities are critical for the final quality of the products and hence in-process targets will be set for each Process control activity and it should be achieved by the shop floor Maintenance/QA team. If it is deviated from targets/standards, machines should not be allowed to continue with production so that there are absolutely `nil` defects in final products –if the system is foolproof.

2.5.3.4 *Jidoka*

JIDOKA facilitates automatic detection of errors/defects in the process and employees in the organization should be motivated to bring out Kaizens focusing on Jidoka. However OEM has a big role to developed Jidoka in their own machinery acting on the feedback from their customers. Examples of Jidoka in Automobile Industry is given below for a better Understanding:

- Seat belt alarm in Cars
- Alarm for Handbrake in Cars

In Spinning mills, the best examples for Jidoka are:

- Alarms in Autoconer Clearers with a provision to stop or eject the cop with defects exceeding the limits
- Alerts on rogue spindles in Ring frame from Individual Spindle monitoring System

2.6 Predictive Maintenance-Third Generation Maintenance

To understand Predictive Maintenance we should first know what are the normal issues related to Maintenance. To put it in simple words:

A. Unplanned downtime and machine failures
B. Low production efficiency
C. Lower level of OEE-Overall Equipment Efficiency(TPM Factor)
D. High spares inventory
E. Higher maintenance costs
F. Poor `***FPY-First Pass Yield***` and Quality metrics
G. Rework caused by `out of spec` machine
H. Poor ***ADD***-Adherence to Despatch & Delivery
I. Inability to identify Root causes of breakdowns
J. Risks associated with Safety of Machines and Employees
K. Risks associated with failures having impact on Environment issues

Predictive Maintenance is a technique that uses tools and techniques to detect anomalies in the operation of machines and possible/probable defects in equipment and process so that we can fix it before `they fail or result in any unsafe event`.

A question may arise on the difference between Preventive Maintenance and Predictive Maintenance which is explained below.

Preventive Maintenance occurs on the same schedule for every cycle. It is based on the assimilated experience of Equipment suppliers, Research Associations and plant technologists.

Predictive Maintenance is proactively figuring out when an asset should be maintained based on its actual state rather than on a fixed schedule. Under Predictive Maintenance, we maintain the machines based on its critical performance metrics to locate the potential problems/failures and address the same beforehand.

Benefits of Predictive Maintenance:

A. Ensures RFTQ-Right First Time Quality
B. Reduces the cost of Maintenance
C. Reduces the downtime of Maintenance
D. Improves the bottom line of the organization
E. Eliminates surprises like sudden breakdowns/sudden rise in replacement costs/sudden increase in overtime expenditure of maintenance crew
F. Has greater impact on the profitability of the company
G. Improves the Strength of Maintenance system as a whole.

PDM Techniques and tools: In predictive Maintenance programme, we have few techniques and tools for enabling the spinners to perform their actions based on `accurate data on the health conditions of the machines`.

A. Infrared thermo graphic analysis/Thermal imagers: With the help of non contact infrared thermo graphic analysis hotspots can be identified in any location of the equipment-accessible or inaccessible. Based on operating requirements of an equipment`s temperature standards are available and hot spots can be identified as the potential spots for immediate correction.
B. Ultrasonic Analysis: These sensors work on the same principle of radar system. It converts electrical energy into Acoustic waves and vice versa too. Typically a microcontroller is used along with ultrasonic sensors for communication.

C. Energy Meters/Load manager: By monitoring the Energy consumed from each machine, we can identify the level of variation between similar machines and take prompt action.

D. AFIS – Testing instrument is used to find out the nep level in the delivered material after beaters/Cards/Combers. By utilizing the trend of neps life span of critical accessories like Beater pins & wires/Card wires/Comber Unicombs and Top combs. By using the resultant value of neps and its trend, the life span of critical items are finalized and replaced. Instead of time based or quantity based replacements in Preventive maintenance systems, PDM helps mills to act scientifically which saves time and money for the mills in a significant manner.

E. Tensojet – Single yarn strength tester is utilized as one of the PDM tool for estimating the weak places in yarn between Ring frames and these data are utilized to decide on replacement of Cots/Aprons/Perforated aprons & inserts in Compact spinning system etc.

Mills should develop a chart on PDM activities for all the critical machines including Humidification plants, Automatic waste evacuation systems, Electrical panels and transformers, Compressor airlines for and inside the machines etc. With the help of Load managers/Energy meters machine wise consumption of energy should be measured/compared against similar machines for taking corrective actions. A model Predictive Maintenance schedule is given in Table 2.7:

2.6.1 Smart Predictive Maintenance –SPDM (Third Generation Maintenance Extension)

SPDM-Smart Predictive Maintenance is a combination of Industry 4.0 and Conventional Predictive Maintenance for the benefit of the industry as a whole. By proper use of permanently installed sensors in critical and vulnerable zones of the Machines/Equipments to measure Heat, Current/energy, Vibration and Air leakage data can be collected 24*7 without any manual intervention.

With properly designed IoT enabled system assimilated data can be analysed and alerts can be generated. Service providers can provide cloud data to benchmark any industry on performance metrics.

Decision support replaces labour intensive data analysis and corrective action with an automated process that identifies the probability of specific faults and prescribes an appropriate action.

TABLE 2.7

Predictive Maintenance Programme for Spinning mills

No	Activity	Frequency	Department/Sections
1	Air leakage analysis with Ultrasonic detector	Once in three months	Complete airline system throughout the loop at vulnerable points
		Once in two months	Autoconers/Lapformers/Cards/Auto Offers in Ring frame
		Once in 6 months	All other machine locations where air is consumed
2	Vibration checking	Once in 3 months	Ring frames/Humidification plant fans/ AWES centrifugal fans
3	Thermography analysis	Once in 2 months	RF motorsHumidification plant motorsAWES Centrifugal fan motorCompressor
		Once in 3 months	TransformersMV panelsVCB/ ACBContactorsRelay switchesCable joints
4	Load studies	Once in 3 months	All the motors above 20 KW in the mills
		Once in an year	All other motors less than 20 KW which is on continuous use

Benefits of SPDM:

A. *Asset monitoring and health*: Real time data on individual assets is available
B. *Condition based Maintenance*: Performance metrics are established and intelligent alerts are generated. Unnecessary stoppages and unnecessary expenditure are eliminated.
C. *Performance based Maintenance:* Performance parameters are tweaked and optimal performance ranges are defined.
D. *Prediction based Maintenance*: Models predict when an asset will fail to operate/function.
E. *Smart Predictive Maintenance*: Maintenance planning is automated and execution is enhanced by Digital Technology.

SPDM Tools in practice in Smart Spinning mills

OEM vendors have identified the need of SPDM tools for the Smart spinners and developed many such tools have been developed by them. Few such tools are explained below:

A. Machinery manufacturers have provided Energy sensors, Air leakage sensors, Vibration sensors, Thermal sensors in the most critical machines and capture real time data. These data are compiled and reported alongwith expected levels of performance. Alarm limits are also set in the machines either to stop or send signals to the concerned responsible persons to take action.

B. Link coner manufacturers have desigined `SP-ID`-Spindle identification system to identify rogue spindles and the cops produced from these spindles are eliminated from production and the report is generated on such rogue spindles to the Spinners for immediate corrective action.

C. Advanced Clearers are provided with Cone expert system to identify `rogue drums` where performance is poor. Machinery manufacturers have developed KPI-Key performance indicators for the performance of their winders and the process. Performance level of each drum against these KPI`s are monitored 24*7 hours and data is converted into reports and suitable alert systems for taking action on rogue drums.

D. TTC-Total Testing Centre from Uster identifies history of Rogue spindles in conjunction with Uster sentinel and UQ4 Clearers.

2.7 Smart Prescriptive Maintenance Framework-SPMF (Fourth generation maintenance)

Smart Prescriptive Maintenance Framework came into existence from the year of 2017. Prescriptive maintenance is the growing technology at a faster pace by utilizing Artificial Intelligence/IoT/Cloud data/Analytics and Machine learning.The information gathered through real time data from the sensors will be utilized to `***Prescribe***`corrective actions and to pass instructions to the users.

Initially this technique has been used by aircraft maintenance team. Later OEM Vendors like Muratec captured the idea of SPMF and utilized the technique for the benefit of Spinning mills. Muratec has developed Visual Manager System on the platform Artificial Intelligence to serve as SPMF.

A. SPMF is identical to SPDM

B. SPMF is a Quantum jump over SPDM to automate the Maintenance practices

C. Over and above SPDM activities, it provides recommendations to `repair or replace components`

Key inputs into the system by Muratec in VMS:

1. Most critical KPI`s of the system like Missplicing %, Upper yarn sensor failure, Autodoffer missing rate, Clearer malfunction drums, Level of tension breaks etc.
2. Knowledge bank-History of failures and corrective actions
3. MTBF-Mean time between failures
4. Remaining useful life of a critical part

Prescriptions:

System is so designed to recommend the users on how to attack the drums where deviation in performance is observed. It also guides on the parts to be replaced and educates the users on the method of replacement with a descriptive diagram.

SPMF is called as SPMF since it is a framework covering:

A. Performance - real time data
B. Benchmarked Performance standards based on cloud data
C. Machine learning-Fault history/Tribal Knowledge/MTBF
D. IoT enablers
E. FMECA-Failure Mode and Effect Criticality analysis

Case studies on Maintenance Management

CS6: Case study on "Impact of Buffing accuracy on Yarn quality"

In a Spinning mill, during the machinery maintenance audit it was identified that the CV of imperfections as very high in Ring Frames as investigated by a random rogue spindle Analysis. As the process parameters and in process performance parameters up to Speed frame were well under control, Buffing schedule and the quality of buffing had been investigated by the Auditor and the findings are given below:

- Grinding stone was not dressed by the buffing operator as per schedule(As per schedule dressing of Grinding stone must be done once after completing two RF sets of top rollers for 1200 spindles)

- Diameter checks after buffing were not carried out by the buffing operator.
- Quality assurance department had not investigated the quality of buffing through U% and evenness studies as per their schedule.

All these deviations were highlighted to the team of Head of Maintenance and Head of Quality Assurance as they are the responsible persons together for ensuring these studies –which are very vital for the Quality of the outgoing product. After discussions with the team, frequency for the dressing stone was set and also SOP was developed to check the `CV of imperfections from online` immediately after buffing. Mills started following SOP and found the CV of imperfections from the Ring frames were under control.

CS7: Case study on "Impact of Air leakage on Mis-splicing% in Autoconers"

In Autoconers, Splicing quality depends on Air quality and Splicer gadgets and settings. In one Spinning mill, we experienced higher Mis-splicing percentage at a level above 2.8% in 60s Combed Hosiery yarn in one Autoconer. Whereas the same count was running in the adjacent Autoconer (same type and model) with less than 1.2% Mis-splicing %. Hence the mills concluded that this was a problem associated with this Autoconer only. Hence mills inspected the same and identified the Air leakage in the incoming distribution air line at several points. Air leakage was also confirmed with an Air leakage detector which is alarming.

Hence Air pipe and joints were totally renewed and Air consumption in the Autoconer was tested while running as well as in idle condition. As it became normal, Autoconer was restarted with restored Splice quality in terms of strength and Splice appearance.

CS 8 Case study on "Process FMEA on Weak places in 80s Combed Compact Warp Yarn"

In one Spinning mill, they experienced higher total weak places in 80s Combed Compact Warp yarn at the level of 45 and the weak places due to strength is more, as tested in Tensojet. The trend is similar in all the four Ring frames running in this Count. Hence the Process FMEA team took up this assignment and the problem was analysed with RCA techniques. Fishbone diagram was used to explain the contributing factors. Through a careful examination and elimination process, the Process FMEA team narrowed down the issue as due to inconsistency in SFC(n) level between Combers which were feeding to these Ring frames.

Quality assurance and Maintenance team had attended the mechanical conditions and settings in the Combers which lead to this higher level of inconsistency in SFC(n) in comber sliver. After resetting in these combers Mills achieved SFC(n) of 5.5 with a range of 0.3 between combers, well under control. Controlled samples were taken in one RF for assessing the improvement and we achieved `0` total weak places after correction of this variation. Bulk level studies were repeated by collecting samples from all the Ring frames which also reconfirmed the controlled study findings.

This experience of Process FMEA had instilled confidence in FMEA techniques in the minds of the cross functional team in the mills effectively.

CS 9 Case study on "Impact of 5 S Implementation in a Spinning mill"

One Spinning mill had decided to start 5S in their organization. An awareness programme was conducted to all the employees in their own language on the benefits of 5S to all the employees in the organization. Similarly a separate programme had also been conducted to all the Shop floor technical staff/ Administrative staff/HR staff and the executives. Once these programmes were completed, a 5S organogram was formed in consultation with the Management. Once the Organogram was in place, the role and responsibilities of Core group members/Team leaders were explained in a separate `focused` training programme. As an usual practice, 5 S was started with `Big Cleaning day`. On `Big cleaning day` an entire team of employees of the organization were entrusted with the job of `SEIRI-Sorting and removing unwanted` in their respective places including environment. After 8 hours of active participation of all the employees, the identified benefits are:

- In every department only the essential items were retained
- In all the cup boards of service rooms, more space was created for further storage in a neat manner

- Stores racks were freed from `obsolete items/shelf life completed items`
- QA department had removed `obsolete yarn/fabric/customer samples` and enough space was created inside the department
- Administration department had thrown out all obsolete records and stationeries.
- Engineering department had segregated Motors above four re-winding and all obsolete items for disposal.
- By disposing of the obsolete items and waste generated during `Big cleaning day` mills had generated an income of Rs 4.5 lakhs which was used by the team for further development works.

Once the shop floor team tasted the benefits of one technique of 5S itself, they started whole hearted committed contributions from the next level of `Fixed Point Photography and Red tag areas`. This mill,with its dedicated 5S team completed 5S implementation within a short span of 6 months itself.

CS 10 Case study on "Design FMEA in a Spinning mill"

In a newly started Spinning mill with new Ring frame, the initial production was started with 40s Combed Hosiery yarn. The production and quality from these products were established for mills standards on Hosiery counts and Customers requirements on Yarn Quality. Six months later mills changed the count in one group of eight Ring frames and were unable to increase speeds. In warp counts production targets could not be achieved if the spindle speeds were not increased to compensate for the increase in the level of twist. Whenever the motor speed touched the Maximum spindle speed of 21000 itself, Motor load would cross standard limits and machine would come to a stop due to relay mechanisms. As the issue was related to machine design, the assignment was taken up with the Design FMEA team. During the discussion with the Design FMEA team, representatives of the Ring frame Vendor also had been included. At that time all machinery parts/settings/ TDS-Technical data sheet had been investigated in depth and team concluded as below:

- As the mills was all along in the Hosiery business, TDS had been filled with data based on 40s Combed Hosiery counts only.

- Maximum speed had been mentioned as 20000 only.
- Hence the Drive wheels/Relay settings/Inverter settings were tuned only for 21000 as a safer limit beyond which machine will be stopped by overload relays.

After these findings, the Machinery manufacturer supplied suitable combination of drive wheels and changed the settings for Relays and Inverter for higher speeds to enable the RF to be run at higher level of Spindle speeds.

Mills were able to achieve higher speeds and were able to achieve standard grams per spindle in 40s Combed warp. Later the machinery supplier provided the same set of wheels for all the other new Ring Frames so that the mills can have flexibility to run on both counts.

The Design FMEA team had also created one SOP for TDS filing to machinery manufacturers from the mills to eliminate such pitfalls by transparent negotiations with machinery suppliers forecasting the future requirements of the mills.

CS 11 Case study on "QAD-Maintenance interface"

In any Spinning mill, there are fewer critical maintenance activities which directly have an impact on Quality of the product as explained elsewhere in this book too.

In a Spinning mill, Card neps were inconsistent immediately after resetting the Card by the maintenance crew. Quality assurance department inspected the in- process results of other parameters in this card and there was no significant deterioration except neps. Hence a thorough re-audit was done on this Card for possible lapses and identified a major crack in the waste collection suction duct from Pre- cleaning zone. The net result of this crack was a drop in suction level which led to accumulation in suction tubes collecting the trash and waste from pre-cleaning zone. After changing the waste collection tube anew, nep level in the card reached standard level with consistency.

CS 12 Case study on "MTTR-Mean Time To Repair(TPM)"

In one of the Spinning mills where TPM was in progress, the Planned maintenance pillar team was analysing the Mean Time To Repair based on the time taken to repair i.e. time taken by different sets of maintenance team to attend different kinds of breakdowns. They found a strange coincidence of one common factor in all the kinds of breakdowns as below:

- Whatever be the type of machinery or type of breakdown, the MTTR in night shifts was nearly 40 to 60% higher than that is experienced in Dayshift.

The team called for a meeting of all the concerned maintenance crew along with Stores in charge and production in charges for identifying the root cause behind the same. After a thorough discussion on several breakdown reports of similar nature, they were able to conclude as below:

- Time taken to open the stores for replacement of parts took invariably ranged from 30 minutes to 120 minutes depending on the availability of Shift Security officer (who patrols entire compound outside factory building in night shifts)
- Conveyance arrangement from the mills for bringing the maintenance crew from their nearby locations was not a structured one which resulted in another delay of 15 minutes to 90 minutes to arrange the conveyance itself.

PM pillar team took the following steps immediately to reduce the time taken on both these `unavoidable extra requirements` in the night shift as below:

1. Keys for the Stores will be kept in time office and the concerned Shift in charge had been authorized to open the stores in case of any requirements.
2. Standard operating procedure had been created for bringing the maintenance crew from their residence without wasting time on red- tapism.

After a period of 6 months, PM pillar reported that MTTR in night shifts was just 10% more time than that was being experienced in the dayshift, which was really a significant improvement against the level of 40 to 60% at the time of MTTR analysis.

Annexure IX A: Model Check list for effective Machinery Maintenance Audit-Online

Machinery Maintenance Audit Check List-off line			
Department/Machines	**Conditions as observed**	**Mills std life**	**Benchmarked mills std life**
Blow room			
Blendomat			
Condition of opening rollers			
Condition of conveyor			
Presence of Seed trap			
Condition of magnets			
Noise level during running			
Line 1 and 2 -Common equipments			
Metal detector operating conditions			
Heavy particle separator operating conditions			
Uniclean -Beater condition			
Uniclean grid bar conditions			
Line 1			
Unimix -Lattice conditions			
-Beater condition			
-Feed rollers			
CVT1 -Beater condition			
-Grid bar free movement			
-Grid bars condition			
Contamination clearers			
-Ejection of valves			
-Illumination lux level			
-Ejections uniformity			
(Repeat 18-27 for other lines)			
Cards			
Cylinder wire condition			
Lickerin wire condition			
Flat tops condition			
Card numbers having cylinder wires beyond life as per mills standards			
Card numbers having lickerin wires beyond life as per mills standards			
Card numbers having flat tops beyond life as per mills standards			

(Continued)

Machinery Maintenance Audit Check List-off line			
Department/Machines	**Conditions as observed**	**Mills std life**	**Benchmarked mills std life**
Card numbers having pre cleaning segments beyond life as per mills standards			
Card numbers having post cleaning segments beyond life as per mills standards			
Card numbers having damaged waste suction hoods			
Auto leveller parts condition			
Condition of the wire mounting equipment			
Condition of the Tools used in Cards			
Pre comber Draw frame			
-Creel drive			
-Draft zone parts			
-Bottom roller trueness(front)			
Lap formers			
-Creel drive			
-Draft zone elements			
-Condition of pneumatics			
-Lap disc clearance			
Combers			
-comber numbers having top combs beyond standard life			
-comber numbers having Unicombs beyond standard life			
-comber numbers having Detaching rollers beyond standard life			
-comber numbers having brushes beyond standard life			
Condition of top rollers			
Top roller loads status			
Detaching rollers condition-Top			
Detaching rollers condition-Bottom			
Coiler calender rollers-condition			
Draw frames			
Creel drive			
Scanning rollers/shaft condition			
Timer belt condition			
Lateral play in bottom rollers			
Bottom rollers eccentricity			
Top rollers condition			

(Continued)

Machinery Maintenance Audit Check List-off line			
Department/Machines	**Conditions as observed**	**Mills std life**	**Benchmarked mills std life**
Top rollers loads uniformity			
Auto leveller condition			
Quality of trumpets			
Speed frames			
Top roller loads			
Toparm setting deviation from standards			
Function of tension control and Bobbin build			
Creel drive			
Buttons in bobbin wheel			
Flyer condition			
Flyer positioning			
Buttons in Bobbin wheel			
Life of top roller shells			
Speed frame numbers having top roller cots beyond std life			
Speed frame numbers having top aprons beyond std life			
Speed frame numbers having bottom aprons beyond std life			
Speed frame numbers having false twisters beyond std life			
Ring frames			
Ring frame numbers having top aprons beyond std life			
Ring frame numbers having bottom aprons beyond std life			
Ring frame numbers having Bobbin holders beyond std life			
Ring frame numbers having Rings beyond std life			
Ring frame numbers having Spindles beyond std life			
Ring frame numbers having Lappets beyond std life			
Ring frame numbers having Compact zone critical elements beyond std life			
Ring frame numbers having Front bottom roller beyond std life			
Draft Zone gear play			
Bottom roller eccentricity			
Presence of clearer rollers for front top rollers			

(*Continued*)

Machinery Maintenance Audit Check List-off line			
Department/Machines	**Conditions as observed**	**Mills std life**	**Benchmarked mills std life**
Clearer roller quality			
Autoconers			
Autoconer numbers having splicer components beyond std life			
Autoconer numbers having Suction combs beyond standard life			

Annexure IX B: Model Check list for effective Machinery Maintenance Audit- Offline

Automatic waste evacuation system
-suction level in Blow room points
-suction level for Lickerin zone
-Suction level for Flat zone
Pre comber drawframe
-Creel drive vibration
-Creel stop motion
-Front stop motion
Lap formers
-Creel drive vibration
-Creel stop motion
-Front stop motion
-Lap initial layer
-Lap density
Combers
-Web appearance
-Suction level
-Sliver visual appearance
Drawframes
Creel drive vibration
Scanning rollers cleanliness
Stop motion-Front
Stop motion-Creel
Speed frames
Performance of the positively driven top clearers

(*Continued*)

Performance of the positively driven bottom clearers
Creel stop motion
Front stop motion
Ring frames
Spindles with vibration(study with stroboscope)
Ringframe numbers having air leakages
Ring frame numbers having vibration
Ring frames having motors with temperature beyond standards
Bottom roller eccentricity
Ring frame numbers having less pneumafil suction level than mills standards
Ring frame numbers having less pneumafil suction level than mills standards
Ring frame numbers having less Compact suction level than mills standards
Autoconers
Autoconers with Air leakages beyond std in idle condition
Autoconers with Air leakages beyond std in running conditions
Autoconers in which repeated Length variation is experienced
Autocners in which cone weight variation is experienced beyond stds

Annexure X: TPM Organogram

CHIEF EXECUTIVE OFFICER

TPM Steering committee

TPM Coordinator

Champion-Development	Champion-Manufacturing Excellence	Champion-Motivation and Morale	Champion-Office Administration
FI & Initial Control	AM, PM & QM	SHE/E&T	OTPM
Shop floor Cells individually for each pillars	Shop floor Cells individually for each pillars	Shop floor Cells individually for each pillars	Shop floor Cells individually for each pillars
Members in each pillar	Every pillar should have minimum 6 members selected from Shop floor machine operatives and Foremen or Supervisors		
Meeting schedule	Pillar cells	Every week at shop floor or in a seminar hall	
	Interaction by Champions for self evaluation	Once in a month	
	TPM Steering committee	Once in 3 months	

(Continued)

Deliverables	One point lessons Kaizen project findings Standard operating procedures Knowledge bank Evaluation of critical performance parameters in line charts or any other suitable means Minutes of meeting with Steering committee

Annexure XI: TPM Gap Analysis Audit format

No	Department	Check points	Auditor`s comments	Grading* (1 to 5)
1	Whole of the organization	1. Availability of KPI		
		2. Availability of Review system		
		3. Adequacy of review system		
		4. Follow up on Management review		
		5. Benchmarking procedures		
2	Shop floor Manufacturing Management	1. Plant Utilization %		
		2. Plant production efficiency		
		3. % of rejected production including seconds		
		4. Overall equipment efficiency		
		5. Availability of other KPI`s		
		6. Practice of Daily meetings		
		7. Availability of action plans with responsibility and targets		
		8. Follow up of action plans		
		9. Breakdown History		
		10. Breakdown analysis-MTBF/MTTR		
		11. Technical data sheet for new products and equipments		

No	Department	Check points	Auditor`s comments	Grading* (1 to 5)
2	Quality Assurance department	1. Availability of Process parameters and cross verification systems		
		2. Prediction of Yarn quality against Customer requirements and cross verification		
		3. Simulation studies for assessment of end use performance		
		4. Customer meetings for designing the products and follow up		
		5. Customer Complaint resolution system		
		6. New Sample development system		
3	Human resources	1. Availability of Induction training manual		
		2. Availability of recruitment and selection procedures		
		3. Availability of Absenteeism control procedures		
		4. Training standards		
		5. Evaluation procedures and records		
		6. Motivation improvement programmes		
		7. Environment maintenance		
		8. Accidents reports		
		9. Safety committee follow up		
		10. Safety system for women employees		
4	Administrative office	1. Filing system		
		2. Office and Factory upkeep		
		3. Security systems		
		4. Stores –inward and outward controls		
		5. Emergency welfare equipments		
		6. Ambulance availability		
		7. First aid equipments		
		8. Fire safety equipments		

3

Production Management in Spinning Mills

LEARNING GOALS

- **Utilization Efficiency**
- **Production Efficiency**
- **Overall Efficiency**
- **Losses on Utilization**
- **Benchmarking Levels**
- **Mill Standard Document**
- **Spin Plan**
- **Production Formulae**
- **Yarn Realization**
- **Invisible Loss**

Introduction

Production management can be termed as efficient only if the product meets the end use performance characteristics and satisfies customers with the effective management of resources. Hence, plant utilization, production efficiency levels, Raw material utilization, Manpower management and energy Management are more important focal points for our attention. Production management is essential in the present day business environment. In a world of global competition, higher production and effective utilization of resources lay the roots for excellence. Measuring our levels of performance and benchmarking ourselves against the peers in the industry for excelling in manufacturing is elaborated in this chapter.

DOI: 10.1201/9781003257677-3

3.1 Utilization Efficiency

Ring frame spindle utilization is the most critical Key Performance Indicator (KPI) influencing the conversion cost. Higher spindle utilization leads to cost reduction per kg of conversion cost drivers such as overheads, depreciation, interest, stores, power and wages. Further it also helps to increase the profit margin due to the higher volume of production and sales turnover. Spindle utilization (SH) is the ratio between the average spindle hours worked per day and the total number of spindles installed. During the process of yarn production in ring frame there are cyclic stoppages due to stoppages for doffing, frequent count/product/mixing changeovers and related setting adjustments. Moreover breakages in the individual spindles of the ring frame leads to idle spindles & reduced spindle utilization of the ring frame affecting the production levels and raising the cost of production. Improvement of ring frame spindle utilization is a vital factor influencing the overall competitiveness of the spinning mill and is necessary in the long term sustainability of the textile industry.

The probable loss due to under utilization of Ring frames in an average mills having 50000 spindles capacity is presented in Table below:

TABLE 3.1

Impact on Net Contribution due to Utilization Loss

Causes	Quantum of loss(in rupees)
1% Loss in utilization in a 50000 spindle mills per day	Rs 17500/-(Approximately)
1% Loss in Utilization in a 50000 spindle mills per year	Rs 63.50 Lakhs(Approximately)

The organization has to utilize the Machines at its maximum operating utilization levels. In Spinning mills, Ring frames are the most important `production machines` and all the other machines (pre and post spinning) are planned to run to meet the requirements of the Ring frames for continuous running. Spin plan is the tool for the spinners to plan Ring Frames with different count patterns with back process machinery and post spinning machinery requirements.It is one of the foremost responsibility of the Manufacturing heads to review the Spin plan against marketing requirements periodically.

Apart from planning from production on requirement of machinery, planned stoppages for Maintenance and QC studies need to be incorporated in our Spin plan.

Factors that should be focused for improving the Utilization of Ring frames are as below:

Controllable factors:

Raw material availability: Raw material requirement needs to be planned once the count pattern is frozen and Cotton has to be purchased in advance. Mills need to take techno-commercial decisions to procure raw material in line with Market demands.

Manpower: Mills need to establish standard practices in employing the operatives for Machines and other employment inside the mills. As far as the engagement of operatives is concerned, mills must have a well defined recruitment cum training plan to ensure that machines are being operated with well trained operatives. Operatives of the machines must be trained in `multi departments` so that it is always easier to shuffle at times of necessity.

Doffing time: Ring frames are bound to be stopped and restarted for every doffing operation. Whether it is a manual doff or it is an Auto-doffing system we must have a measurement and control system for controlling the doffing time even though root causes are quite different. As doff time is a cyclical event the impact of the same on Ring frame utilization is significant.

Lot/Count changes: Mills need to stop the Ring frames and make count or lot changes whenever new count/product/mixing is introduced. Mills must have a well defined Lot/Count changes procedure so that the time taken is the least at Ring frame by effective coordination of personnel concerned with. Also mills must avoid frequent lot and count changes which are possible only by a better Raw material procurement system in coordination with the Marketing department.

Planned maintenance stoppages: For routine and preventive Maintenance of Ring frames, we need to allow suitable allowance in our Spin plan itself. It is the responsibility of the Head of Production to ensure these time limits are always followed.

Breakdowns: Breakdowns are not `unavoidable` if there is a proper system of Maintenance. Mills must ensure that their Maintenance system is foolproof. In the case of any such breakdowns, there should be a well structured `breakdown analysis` to eliminate recurrence. Fire accidents too need to be treated similarly.

Labour unrest: Mills cannot afford to have Utilization loss due to any labour unrest on issues related to shop floor or their welfare or any other organizational decisions. Through well designed HR policies to motivate employees their morale can be kept at an elevated level to improve their contribution to the organization. The employees should be well aware of the losses due to utilization for the organization and in turn to them.

Uncontrollable factors:

External failures like Power failure, Harthal, Break out of pandemic issues, natural calamities like floods etc could lead to loss of Utilization which sometimes go beyond the scope of the organization control.

Achievable Levels of Utilization in Spinning Mills

Today the best Spinning mills achieve 98% and above and few top ranking mills are able to maintain consistently above 99%. In the recent study conducted by SITRA, top 20% of the mills had achieved 95.7% Utilization and the best mills have crossed 99.2% Utilization. Mills which outperform others in Ring frame Utilization follow TPM techniques/ FMEA/RCA in Shop floor management effectively.

RCA needs to be followed for every breakdown in Ring frames and also for analysing the `breakup of utilization loss` on a periodical basis. The breakup of utilization loss in the best mills is given below to benchmark the utilization levels on Ring frames.

TABLE 3.2

Breakup of Utilization Loss in Ring Frame

Losses due to	%	Losses due to	%
Maintenance	0.5	Total	0.8
Mechanical breakdown	0.1	Achievable Utilization efficiency(100-total loss)	99.2
Electrical breakdown	0.1		
Traveller changes	0.1		

The loss in utilization on every day needs to be analyzed by shop floor personnel and necessary corrective actions have to be taken on the subsequent day. It is a cyclical process and improvements should be reviewed by the top management with the team periodically.

3.2 Production Efficiency

Spinning mills design their layout and machinery for a defined count pattern with a wide range of production flexibility. When the market requirements are finalized by the top management, the Production plan will be estimated which will cover:

- Cotton requirement
- Spin plan (**Model Spin plan is given in** Annexure XII A & **Production formulae used in Spin plan are given in** Annexure XII B)
- Estimated Production Statement

Spin plan is the basic document for any Production head and the commitments given to the top management through EPS (Estimated Production Statement) must be realized by implementing necessary process changes at the shop floor with the team.

In Spin plan three factors play a crucial role:

- Utilization of Ring frame and all other necessary machines
- Production efficiency of Ring frame and all other necessary machines
- Yarn realization

Utilization has been discussed in detail in 3.1. By production efficiency we mean the efficiency level at which we achieve the target output i.e.

Production efficiency = (Achieved production in kg/Target production in kg)*100

In spinning mills, we will be running many different counts and the production efficiency of the Mills is a weighted average of production efficiency in each count.

Production efficiency in each count = (achieved grams per spindle per 8 hours/target grams per spindle per 8 hours)*100

Production target in grams per spindle for each count needs to be fixed judiciously by the mills by considering the following factors:

- Raw material type and quality
- Age of the Ring frames
- Level of modernization and up gradation in Ring frames and critical machines in preparatory and post spinning
- Level of Automation
- Use of improved accessories or gadgets(e.g) Slim tubes, Lower Ring diameter etc
- Yarn quality requirements by the customers
- Benchmarked standards by Research associations

Mills should maintain `Targets for every variety of products it intends to spin` and this document needs to be upwardly revised periodically with the intervention of the top management and the technical team. Model Mills standard document for spindle production per 8 hours for a Spinning mill is enclosed in Annexure XIII

To achieve higher production efficiency in Ring frames, we need to focus on:

- Idle spindles
- Pneumafil waste

Mills must have a strict control on idle spindles by snap surveys in Shifts and RCA on snap surveys helps the shop floor team to continuously improve and maintain better than standard levels always.

With respect to Pneumafil waste, mills must have a `Machine wise/ Operator wise assessment and control` without which this cannot be precisely controlled. Higher the pneumafil waste, higher will be the deterioration in yarn quality as mills cannot afford to sell the pneumafil waste but it can be utilized only in mixing.

Probable causes for higher Pneumafil waste are:

- Over spinning
- Traveller selection
- Speed pattern
- Operating temperature and Relative humidity
- Drafting parameters like settings/draft/spacers
- Condition of critical accessories like Rings/Spindles/Cots/ Bobbin holders
- Efficiency of the operator on Patrolling time/mending time/ piecing efficiency
- Quality of the Bobbins
- Quality of the compact accessories in case of Compact spinning including perforated aprons and inserts

Once every shift is completed, kg produced in each Ring frame and also the grams per spindle achieved in each Ring frame for the `running hours` will be entered in the production reports. Production efficiency will be estimated for each count.

As an example if the mills has achieved 48 grams per spindle for 8 hours and the mills target is 49.5 for 80s Combed weaving count then the production efficiency for 80s Combed weaving is 97% i.e.{(48/49.5)*100}

Production efficiency of the shed for the particular shift will be the weighted average of production efficiency of all counts for the running hours in each count.

For example if the mills has achieved 97% in 80s Combed weaving with worked spindles of 20000 and 90% in 30s Combed hosiery with worked spindles of 30000 then:

$$\text{Production efficiency of the shed} = \{(20000 * 0.97 + 30000 * 0.9)/50000\}$$
$$= 92.8\%$$

To enable benchmarking ourselves against top ranking mills, SITRA has developed `40s Adjusted conversion factors` which can be used to assess us against Benchmarked mills. As an example the following table has a few examples for `40s Adjusted conversion factors` for a few selected counts.

TABLE 3.3

Conversion Factors * for 40s as per SITRA`s Revised Productivity Standards (2019)

Count	40s Adj Conversion factor	Count	40s Adj Conversion factor
20s CH **	0.332	20s CW ***	0.339
30s CH	0.535	30s CW	0.553
40s CH	0.752	40s CW	0.846
50s CH	1.045	50s CW	1.176
60s CH	1.374	60s CW	1.623
80s CH	–	80s CW	2.499
100s CH	–	100s CW	3.773

Notes:

*SITRA has established norms for 40s Converted production as 116 grams based on the latest surveys it has conducted among its leading member mills.

**CH-Combed Hosiery

***CW-Combed Weaving

3.3 Benchmarking

Spinning mills need to benchmark their units` performance against Industry standards. Primarily in India, SITRA`s production standards are used to benchmark any Spinning mills. Let us see the same with few examples:

Example 1

Let us assume our mill has achieved 44 grams per spindle per 8 hours in 80s Combed weaving yarn in Ring frame. We need to find out what is the equivalent of 40s adjusted production. As per the above table, the conversion factor for 80s Combed weaving is 2.499.

Hence 40s Adjusted grams for our mills level of 48 grams $= 44 * 2.499$
$= 110$

SITRA Standard for 40s converted production in gpss $= 116$

Hence our level of efficiency against benchmark is $= (110/116) * 100$
$= 94.8\%$

It can be concluded that the mill has a scope of 5.2% against SITRA standards and should try to improve further to reach SITRA standards.

Example 2

Let us assume a mill achieved 202 grams in 30s Combed hosiery yarn in Ring frame. If we commute in the same manner as above our level of efficiency against benchmark in 30s Combed hosiery will be 93.1%.

Mills has a scope of further improvement of 6.9% to reach the benchmarked levels.

Practices to be followed to achieve higher production per spindle:

Mills need to establish their systems and procedures to achieve higher grams per spindle and sustain by continuous in house brainstorming sessions. Few systems are discussed below:

- **Raw material Selection**: Right kind of raw material to meet the Yarn quality requirements of the customers and is essential. However raw material selection calls for acumen to decide the running performance of the raw material which decides the production efficiency at shop floor.
- **Individual spindle monitoring systems**: These are the latest devices introduced in Spinning mills. Sensors are incorporated against each ring to track the movement of travelers from which the necessary data are retrieved about:

- Idle spindle
- Idle time
- Rogue spindles
- Doff time
- End mending time
- Production per spindle
- Pneumafil waste

These systems enable taylor made report formats also to suit shop floor Management as well as Top management. By utilizing these reports timely, for performance reviews and initiating suitable actions at all levels will help the mills to improve grams per spindle competitively.

- **TPM implementation**: TPM needs to be practiced by the mills which enable the mills to do RCA for every drop in spindle point production and to introduce innovative practices for improving the production continuously. TPM helps the mills to take right kind of actions in time to achieve the best levels of machine utilization and high levels of production and effective utilization of resources.
- **Training the Production operatives and Shop floor employees:** Mills need to follow scientific methods of training, review and upgrade their training standards to scale up the performance of their employees to achieve higher levels of production. Shop floor technical executives need to be well trained on Machines/ Technology/Process and their improvements are also monitored for scaling up their contribution. As this part of activity is one of the essential tool to achieve manufacturing excellence a separate chapter is devoted on `Training` where we have covered the ingredients of training with a focused approach.

3.4 Yarn Realization

Yarn realization is defined as the ratio of quantity of yarn produced to the quantity of Raw material consumed. (i.c.)

Yarn Realization =

(Total yarn produced in kg/Total quantity of cotton used to produce the above quantity of yarn in kg)*100

For example if 7200 kg of yarn is produced from 10000 kg of Cotton, we conclude Yarn realization as 72%. Balance 28% is termed as saleable waste and invisible loss.

Mills need to control Yarn realization as it is one of the prime factors controlling the cost of manufacturing a product. When the yarn realization is lower than the Mills standards then the cost of mixing will escalate as the manufacturing cost of yarn is arrived at by estimating the clean cotton cost.

Clean cotton cost =

{(Cost of mixing/Yarn realization)-selling cost of waste generated}

If clean cotton cost is increased then the profit per kg of yarn will be affected as Profit per kg of yarn = Selling price of yarn-Manufacturing cost including clean cotton cost.

Hence, to control the Cost of Manufacturing we need to primarily control Yarn realization in the mills which can be judiciously followed up from Raw material to shop floor.

Breakup of Yarn Realization

Yarn realization is primarily dependent on the trash content in the mixing issued for a particular product group apart from specific value additions in Combing for every product depending upon customer's specific requirements.As per the industry`s established practices,

$$\text{Expected Yarn Realisation YR} = (100 - 3.5T) * 0.97 \text{ for all carded counts \&}$$

$$= (100 - 3.2T) * 0.97 * (1 - C/100)$$

$$\text{for all combed counts}$$

Where `T `stands for Trash content in the mixing (weighted average)&

C stands for weighted average of the comber noil for all the products in that mixing.

When a mixing is laid and processed, we extract waste at the Blow room, Cards and combers which can be set for the particular count group as per our customer's requirements. Over and above there will be one more waste in the Autoconer department as hard waste (yarn waste). In any mills apart from these wastes there will be other category of wastes like Sweeping waste, Filter waste from AWES and Humidification plants (including exhaust trench waste) and Micro dust from filters. Hence, Yarn realization is arrived at mills every month by taking process stock by 1st of every month. A Model worksheet for Estimation of Yarn Realization is given in Annexure XIV. Waste collected in each department will be expressed as a percentage of waste to mixing issued to arrive at the clean

TABLE 3.4

Estimation of Yarn Realization

Category of waste	Production in each department	% of waste on debt. production	% of waste on Mixing issued
Blow room waste	100	3.5	3.5
Carding	96.5	7.5	7.2375
Combers	89	16	14.24
Filter waste and sweepings*	73	2	1.46
Hard waste	73	0.5	0.365
Total saleable waste(SW)			26.8
Expected Yarn Realization (100-SW)%			73.2
Actual Yarn Realization		Expected YR-*Process invisible loss*	

Note:
* Here the filter waste and Hard waste are accounted for on Spinning production.

cotton cost. An example for estimation of Yarn realization based on waste level extraction in each department is given in Table 3.4:

3.4.1 Invisible loss

Invisible loss is the difference between actual mixing issued minus the total sum of yarn produced and waste produced. Even though, it should be equal, there is bound to be a difference to the tune of 0.5% to 1.5% depending on the Factors below:

- Inconsistency in Mixing quality with respect to micro dust level
- Moisture content in the bales received
- Moisture content in the wastes sold
- Moisture content in the outgoing yarn
- Accounting methods during process stock and its pobable inconsistency
- Accuracy in weighments and the calibration status of weighing machines
- Actual weight of yarn dispatched vs invoice weight

Mills need to establish SOP for accurate weighment of everything whether it is cotton or waste or yarn so that the errors are minimized. Similarly we need to measure the moisture content in all these materials and maintain standard conditions. Maintaining the lower invisible loss in the process helps indirectly on improvement of Yarn realization and also to plug the drain in income.

3.4.2 Best Practices To Achieve Higher And Consistent Yarn Realization & Control On Saleable Waste

Raw material selection

Raw material must be chosen with least trash and micro dust. Moisture content in the bales should not exceed 8%. If any lot has more than 8%, then vendors must be suitably charged for the losses incurred by the mills.

Mills must have the system of individual bale weighment for effective control on mixing input. Rather mills can follow a statistical random weighment of individual bale weights to arrive at `quantity of mixing issued`.

Settings in Blow Room and Cards

The waste levels in Blow room and Cards should be set as per the `in process quality requirements` for the particular product. Once it is set, zone-wise analysis of waste must be done periodically so as to maintain the waste level at each beating points/cleaning points/cards. Model zone-wise waste analysis chart for Cards is enclosed in Annexure XV.

Apart from this selective QA studies based controls, Production executives should check the wastes collected in every shift/day at AWES (Automatic Waste Evacuation System).This assessment is very critical and alerts the Management on any kind of lint loss in the department as a whole. This helps as a tool for continuous improvement zones for the mills with respect to Yarn realization.

Department relative humidity (RH) and temperature plays a major role in Yarn realization as it could lead to fly generation if hot conditions prevail in the department.

Department wise waste levels need to be fixed and followed up. As general guidelines the following industrial norms can be utilized:

TABLE 3.5

Industry Norms for Saleable Waste Extraction

Department	Total waste	Remarks
Blow room	0.75*T	For Indian cotton
	0.5*T	For imported ELS cotton
Cards	2.5 to 2.75 *T	For Indian cotton
	2.7 to 2.9*T	For Imported ELS cotton
Comber noil	8 to 24% On comber production	Primary governing factors are Customer`s specific yarn quality requirements & techno commercial strategy

In imported ELS cotton the trash is normally very less and it has fragmented leaf and seed particles which can be removed only in Cards preferentially. Hence, we must carefully decide the settings in Mote knives and Stationary flats above lickerin/Trash knives for gentle removal of fragmented leaves and seeds at Cards.

Based on the above norms and in process quality standards on Neps in Blow room and Cards mills must further fine tune the zone-wise waste level to achieve desired results.

3.4.3 Measures to Achieve Standard Yarn Realization

Once the Standard yarn realization is finalized based on sampling techniques and techno commercial inputs, mills must take efforts to achieve this STD YR similar to Utilization and Production efficiency. In a medium count spinning mill with 30000 spindles the losses per month due to Yarn realization will be approximately Rs 5 lakhs for every 1% drop in case of Indian cotton. Hence, it is imperative we need to stay focused and take efforts in the following direction:

- Proper weighment system upon receipt of lot arrival and also during mixing issue
- Moisture content measurement and suitable remedial actions from Raw material to packing godown
- Use of fog masters in Blow room/mixing room depending on external weather conditions and the facilities available in Humidification plants of the mills
- Assess the consistency in trash content level in lots used and take suitable corrective actions in process
- Estimate the bulk level waste generation in AWES on an everyday basis and take suitable shop floor actions to eliminate lint loss if any

- Establish control on fly liberation by assessing the level of plant waste in quantity as well as quality, through visual observation.
- As moisture loss is inevitable in the Spinning process, condition the yarn before packing *even if there is no specific request* from the customer.
- Periodical assessment of Waste in Blow room and Cards—zone wise and take necessary corrective action
- If the yarn is conditioned one, ensure the consistency of Moisture gain between the lots
- Proper weighment system for lot despatches and strict adherence to tolerance limits
- Use well trained WIP assessors during every month stock taking for eliminating errors in estimation of WIP
- Maintaining the WIP well within the industry norms of maximum 7 days production even in the mills running fine counts, at any point of time.

3.5 Manufacturing of Sustainable Products

Spinning mills can contribute to the society through manufacturing of sustainable products that save `Mother Earth` and help to save the future generation. Of these many innovations on sustainable production, Organic cotton cultivation across the globe has gained importance and in India also the share of Organic cotton production forms 2% of the total production of total cotton production in India in the year 2020. (However the share of India`s Organic cotton production on global organic cotton production is the highest at 51%.) The percentage of organic cotton yarn production in Indian Spinning mills is also in the steady growth trajectory due to demand in hygiene and medical textiles. The consumers in the evolved markets now prefer to use products that are manufactured with naturally available raw materials that are free from artificial ingredients or grown without chemical additives-these are broadly categorized as 'organic products'. Acting on this significant consumer trend, Mills are using all their resources towards developing natural products, organic raw-materials, and eco-friendly processes. It's no surprise, then, that the Textile Industry is rapidly seeing an increase in the demand for Organic clothing.

Textiles made of Organic cotton is particularly in high demand and replacing the use of conventional cotton. The world is shifting to Organic cotton as it requires much less water and isn't treated with any kind of genetically modified seeds or pesticides. According to studies, India is the largest producer of Organic cotton, with approximately 51% of global supply, followed by China (19%), Kyrgyzstan (7%), Turkey (7%) and Tajikistan (5%). The global market size of Organic cotton crossed USD 37 bn. in 2018, and growing at a Compound Annual Growth Rate (CAGR) of 9.2%. And yet, the entire Organic cotton comprises of only 1.1% of the total cotton production of the world. This is a great indicator of the scope of growth for Organic textile industry.

The Indian textile industry is making a mark in capitalizing on this opportunity while following international standard regulations. Many standardizations and accreditations have been set up to ensure the promotion of organic yarn. The certifications and accreditation for manufacturing organic yarn in the mills are `Global Organic Textiles Standardization (GOTS)& `Organic Content Standards (OCS-IN)`.

Each of these certification bodies have set up certain regulations for textiles producers in the value chain to comply with. In the case of Global Organic Textile Standard (GOTS), only textile products that contain a minimum of 70% organic fibres can become GOTS certified.

Products certified to the Organic Content Standards (OCS) may use the terms "Made with X% Organically Grown Material" or "Contains X% Organically Grown Material," and make reference to the OCS. 'X' must represent the final percentage of Organic Material by appropriate unit of measure in the finished product; 'Material' refers to the actual organic input (e.g. cotton, silk, linen etc.)

In the long-run demand for Organic clothing is expected to grow, therefore developing the Organic yarn production in the mills would be a worthwhile opportunity as a means of profitable diversification. India has a much larger role to play to capitalize on this demand, with cotton being one of its widely-grown crops. Next to Organic cotton we have blends with Natural fibers like Hemp, Linen, Jute and Bamboo.

3.6 Manufacturing of Value-Added Yarns

IND-RA report (Indian Ratings and Research Reports) based on their recent study says clearly that "Large textile houses have focused on Value addition and diversification which helps them to continue sound operational performance".To enhance the profitability of the Spinning mills, mills must explore all possible avenues for value addition in their organization by modernization/upgradation. Some of the successful avenues are as below:

- Ultra contamination controlled yarn
- Polyester cotton Blends
- Modal/Cotton blends
- Bamboo/Cotton blends
- Flax/Cotton Blends
- Fancy Yarn(Slub yarn/Neppy yarn/Vario Siro/JASPE yarn etc)
- Funcional yarns(Anti microbial/Anti bacterial etc)
- Melange yarn(Grey/Dyed/Compact Melange)
- Innovative and Eco friendly yarns
- Recycled yarn

Value addition of yarn helps the Spinning mills to improve their net contribution per Ringframe by a minimum of 20% as the sale value of value added yarns fetches an attractive premium over normal yarns. This has been substantiated by yearly surveys conducted by SITRA on the profitability of the Spinning mills.

3.7 Automation in Spinning Mills

Manufacturing in Spinning mills is most likely to transform in the coming years towards a fully automated operation in the coming years. Currently mills are witnessing a shift from human dependence to automation.Most of the operations in the spinning mills are being automated so as to eliminate human intervention in operation or to reduce it significantly. Shift towards automation can be attributed to the rising adoption of industrial robots in the manufacturing activities apart from the extensive use of Artificial Intelligence(AI). We divide the automation in spinning mills into three parts to have a clear focus and understanding.

3.7.1 Automation towards Yarn Quality Improvements

Meeting the customer requirements is the topmost priority assignment for any spinning mills.In this regards, we have elaborated ZDQC concept in clear terms in Chapter 2. ZDQC calls for efforts in in-process to "Build the Quality into the products". However with normal stereo typed Quality Assurance/Control activities,we cannot assure 100% checking of process/ products. Automation in Quality Management helps in continuous online monitoring of process/products for measuring the deviations and systems are in-built to:

A. Stop the machine/parts of the machine in case of deviation if deviations beyond acceptable tolerance limits occur(e.g) Stopping the Drums in Autoconers if more tension breaks are observed.
B. Correct the deviations and continue the production without defects(e.g) Autolevellers in Cards and Drawframes
C. Reject the defective products and continue the production(e.g) Defective cops in Autoconers are ejected and directed to a separate designated collection point and the production in the particular drum is continued without stoppage.

3.7.2 Automation toward Production Improvement

The objective of automation toward production improvement is three fold:

1. Eliminates or reduces the downtime of machine or its parts significantly(e.g) Autoconers are provided with Auto speed control systems in Link coners to increase/adjust speeds to exhaust the cops within the doff duration of Ring frame.
2. Continuously look for production improvement and alerts the responsible persons for taking actions to improve or automatically adjust the production speeds to improve production from the machines(e.g) Individual Spindle Monitoring systems which are aggressively being implemented by Smart spinning mills to improve production and productivity.
3. Continuously monitors the waste levels ejected from the machine for Quality/Quantity and take necessary actions automatically to reduce waste level(e.g.) Modern cards are equipped with sensors to monitor the lint level in trash ejected out from Lickerin and adjust the settings automatically to reduce the lint levels.

Through the help of such devices in the spinning mills Production improvement is achieved by the spinners effectively through automation. Apart from these devices it is necessary for the Spinning mills to automate the `estimation of production and utilization efficiency of the equipments` so that the human errors and their intervention is totally eliminated. All new machines are being equipped with advanced technology for accurate production monitoring systems which keep record of all stoppages/all idle spindles/idle drums/waste generation etc so that the production estimates are accurate and precise. Moreover these systems give timely warnings to the responsible executives on machines with lower efficiency in time so that actions can be taken immediately to rectify and restore to normal production.

3.7.3 Automation toward Improving Labor Productivity

In a Spinning mills, our final goal on Automation is to convert the raw material available at one end and to deliver the finished yarn in pallets/ cartons/bags so that it can be transported to the customers without any human intervention. However spinning mills have several stages of processing which makes it a herculean task for Automation. In the last two decades, many Automation in spinning mills help to eliminate the use of manpower or to reduce the same significantly. Key automation areas which are already implemented by the Spinning mills are:

A. Automatic Waste Evacuation System(AWES): This system collects the waste from Blow room beaters/openers/cleaners,Cards and Combers automatically through the use of effective suction generated by Centrifugal fans.
B. Automatic Waste Baling press: All the saleable wastes collected by AWES and all the usable waste generated from the in process machinery by Centralized waste collection systems are automatically transported and baled categorywise.
C. Automatic Blending and Mixing: Stack mixing is totally eliminated in 95% of the Spinning mills with the introduction of Automatic and high production blending machines like Blendomat/Bale plucker/Unifloc are introduced. These machines operate at higher production levels too and handles upto four varieties of mixing at the same time.
D. Auto feed to Cards: Materials from Blow room is delivered to Cards directly through chute feed systems with regulated feed control systems continuously.
E. Contamination clearers in Blow room: One of the best development in Spinning mills is the development of Contamination clearers. These Clearers help to identify all types of contaminations in the raw material tufts at a material speed upto 25 m/sec and eject it from the running material. Contaminants ejected by the devices are:
   1. Colour fibers/fabrics/yarns/papers(Both dark and light)
   2. Leafy matters/Heavy trash/Yellow colured cotton/Oil stained material

3. Glittering material like Polypropylene and its equivalent fibers/fabrics
4. Non glittering material like Dull Polypropylene and other contaminants resembling the colour of cotton

The contamination clearers use Monochrome cameras, Colour (RGB) cameras, Sonic sensors, UV lights and Spectroscopic sensors supported by advanced AI technologies. Depending on the rising or lowering level of contaminations in the running material the settings are periodically adjusted automatically. A spinning mill with 25000 spindles manufacturing medium count yarn have a scope of reducing a minimum of 75 employees engaged in contamination cleaning operations in blow room per day.

F. Integrated Draw frame with Cards: Fifth generation Cards are linked (optionally) with Draw frames directly and this eliminates the manpower required for material handling and reduces the manpower required for Cards and Draw frames as both these machines are integrated.

G. Automatic Lap feeding in Combers: Modern Combers have the provision of Automatic lap feeding and auto piecing arrangements at the time of lap exhaust. This automation significantly reduces the requirement of manpower in Combers.

H. Auto doffers in Speed frames/Ring frames/Autoconers: Doffing operation in these machines call for skilled employees and the time taken for doffing also affects the Utilization of these machines. As an example, doffing operation in a Speed frame with 220 spindles take a minimum duration of 15 minutes and doffing in a Ring frame with 1824 spindles take 3.5 minutes. With the help of Auto doffers in these machines doff time is reduced to 2.5 minutes and 1.8 minutes respectively. Apart from the same, no manpower is necessary with this kind of Automation. Similarly, waiting time of drums in Autoconers is eliminated by the use of Auto doffers and it increases the utilization of drums effectively for production.

I. Bobbin Transport Systems: Once the roving bobbins are auto-doffed, these bobbins are either stored in storage areas or transported to the respective Ring frames. Many flexible customer centric taylor made designs are developed to meet the diversified needs of the Spinning mills as below:

1. Closed loop system where bobbns from one speedframe will be linked to a fixed number of Ring frames (suitable for mills running several mixing varieties and fixed number of Ring frames for each variety)
2. Random and Auto feeding system: In this system Bobbins from any group of speedframe can be fed to designated group of Ring frame in the Shed depending on the need randomly. (Useful for the mills where multiple varieties of mixing and multiple counts are being run)
3. Auto creel Changer system: In this sytem the entire creel of Ring frame is replaced by bobbin trains only instead of fixed creels. Whenever the bobbins are exhausted the entire set of bobbin trains will move out and a fresh set of Bobbin trains will be moved in and the process will be continued.

J. Link coners: Link coners collect cops continuously from Ring frame at the time of doffing and feed to the Link coner running drums. Linkcones have a lot of flexibility with it advanced technology with a scope of reducing manpower significantly. In a medium count Spinning mill of 25000 spindles we have a scope of reducing 24 employees per day with the introduction of linkconers.

K. Automatic Packing System: Mills pack the final yarn in different forms depending on the customer's specific requirements and also the destinations. Some common forms are:

1. Bag packing
2. Carton packing
3. Pallet packing

Apart from the above variants, mills follow yarn conditioning for the yarn depending on the type of product/markets.

Automatic packing systems are developed by the manufacturers meeting all these requirements alongwith a provision for online checking of cones for visible package defects including shade variation. Cones which are picked out from the Link coner conveyor pass through UV sensor cabin (fromwhere bobbins with shade variation are diverted and packed separately) are arranged in pallets for conditioning. After conditioning it will be conveyed to Automatic Bag/Carton/pallet packing systems. If yarn conditioning is not required, it can be byepassed too. The entire automatic packing system can be managed by a single employee per shift for a spinning mill having 25000 spindles running on medium counts.

Case Studies on Production Management

CS 13 Case study on "Spindle point production improvement in 80s Combed Compact weaving"

In one of the Spinning mills, 80s CW Compact count was running in a Modern Ring frame. The Ring frame process parameters are given below:

Count: 100s CWC
TM /tpi: 4.05/40.5
Ring diameter and spindle lift: 36 mm/160 mm
ABC Ring diameter: 39 mm
Average spindle speed: 21250 rpm
Traveller used: 20/0 EL1UDR
Spindle point production (grams per spindle/8 hours) = 36.5

While benchmarking process mills identified that there is a scope to increase Spindle point production to above 36.5 by a minimum of 4% while comparing with its peers in the industry for the equivalent counts.

Production team had a brainstorming session with a cross functional team and analyzed the process parameters in detail. Their conclusion based on their brainstorming session was as follows:

- With the current process parameters and the technical specifications of the Ring Frame, the motor was running at its 90% load at maximum spindle speed of the Ring frame.
- With stroboscope analysis, the balloon shape throughout the build had been inspected and found it was well under control at these existing parameters. If the speed was increased beyond it, there was a probability that the balloon will collapse leading to end breaks.As the current level of end breaks per 100 spindle hours was <4, Spindle speed was not able to be increased beyond this level.
- However as the mills used to spin only 80s/100s/120s & above, there was a scope of matching the Ring diameter to finer counts. As the mill was using only 36 mm, it was

decided to take a controlled trial in one Ring Frame with 34 mm.

- Mills procured 34 mm Rings and associated ABC rings with a diameter of 37 mm(Ring dia plus 3 mm) and incorporated in one Ring frame. Controlled studies had been taken with these 34 mm Rings and all the Yarn quality parameters were also assessed alongwith End breaks per 100 spindle hours and pneumafil waste.

Trial findings:

- Mills was able to increase the maximum spindle speed to 22500 rpm with a lesser load on motor at maximum speed
- Balloon shape was studied with a stroboscope and it is perfect throughout the build.
- End Breaks were found to be less than 4 and the Quality assurance department confirmed that there is a specific quality improvement in terms of Uster Yarn Hairiness index by 0.2 consistently.
- Mills was able to achieve 38.5 Grams per spindle which is more than 6.5% over the existing Spindle point production.
- Based on this successful trial, Management had decided to change the Rings in all the 30000 spindles in the particular shed for fine counts.
- As per the payback sheet, payback for the rings along with ABC Rings was less than 6 months.

CS 14 Case study on "Impact of Short stretch in Compact Spinning Ring Frame"

One Spinning mill having 20000 Spindles was manufacturing counts above 60s Combed weaving yarn. All the Ring frames were new and all were with Long stretch only. As the mills decided to go in for Retrofitting of Suction Compact Spinning system, modification of the drafting system had been done in one Ring frame. Mills decided to produce 60s Combed weaving yarn with an improvement in yarn quality in terms of Hairiness but at a higher level of spindle point production. In this Ring frame, mills was able to achieve 10% higher spindle point production in comparison to Normal 60s CW. The trial findings in Long stretch Compact system was as below:

- Spindle point production was increased from 66 to 72.6 grams per 8 hours/spindle
- Imperfections were reduced by 10%
- Yarn hairiness was reduced by 35%

However this production level is lower than the benchmarked level of 75 grams by the mills against their peers in the industry. Hence, the production team had investigated further scope of improvements in spindle point production.

Based on the brainstorming session by the technical team in the mills it was planned to retrofit the Suction compact spinning along with Short stretch conversion(technically, geometry conversion) in the second Ring frame. In the Short stretch Ring frame Spinning geometry was being changed to rectify the changes in the Spinning angle after retrofitting the Compact spinning system. Production studies in the first(long stretch compact) and the second Ring frame (Short stretch compact) was conducted as bulk level studies and the mills concluded as below:

- Spindle point production–77 grams per spindle/8 hours was achieved for equivalent quality level and end breaks as compared to the first Ring frame.
- Mills also found out that there was a consistency in the Hairiness levels in the yarn proved by its lower level of CV of Yarn Hairiness.

As a conclusion of this case study, Management decided to convert all the Ring frames from Long stretch to Short stretch while retrofitting Compact spinning system in their Ring frames to realize the real benefit.

CS15 Case study on "Improvement of Spindle point production with Slim tubes"

Spinning mills normally use Ring tubes with a thickness ranging from 2.4 mm to 3.2 mm with so many permutations and combinations of requirements. However for normal as well as Compact Hosiery and Warp counts, developments had taken place on `Slim tubes` i.e. which are aimed at improvement in Cop content. A dedicated trial had been conducted in one Spinning mill producing 30s Combed Hosiery counts in the latest generation Ring frame with Normal Ring tubes having a thickness of 2.4 mm and Slim tubes having a thickness of 1.8 mm. In this controlled study we observed the following improvements:

Count: 30s Combed Hosiery

Ring frame: Latest generation RF with 170 mm lift(190 mm tube length)with a Ring diameter of 36 mm and cop diameter of 34.5mm and a cop content of 46 grams.

Case study findings:

1. Cop content had gone up by 12% i.e. from 46 grams to 51.5 grams
2. Due to the Speed pattern improvement and available extra time for cop build the average speed had gone up by 400 rpm which resulted in enhancement of Spindle point production by 2%
3. As the cop content was increased by 12%, the number of cop changes was reduced in Autoconer which increased the Machine efficiency by 3.5%

As a conclusion to improve production level, the mills decided to replace the Ring frame tubes in the entire mills within two years with a strategy based on the life of the existing Ring tubes.

CS 16 Case study on "Utilization efficiency improvement"

In a particular Spinning mills, the Average Utilization efficiency was consistently maintained at 97.9 % over a 6 month period. However the benchmarked level against the Spinning mills was 99.2%. Hence, the Technical team from Production/Maintenance and Engineering divisions joined together and analyzed using RCA(Root cause analysis) technique. The findings of the meeting were:

1. There is a definite scope of 1.4% between the benchmarked Utilization and the mills as per average Utilization efficiency
2. Break up of Losses in Utilization efficiency was arrived at as below:
   - Maintenance - 0.7%
   - Mechanical breakdowns - 0.4%
   - Electrical breakdowns - 0.25%
   - Traveller changes - 0.18%
   - Creel changes - 0.20%
   - Back stuff shortage due to Lap former breakdowns - 0.4%

1. Maintenance team had been asked to study their maintenance schedules and to be extended without affecting the quality of the product, by resorting to `online cleaning of Ring Frames`. Target for Maintenance was revised to 0.5%
2. Mechanical and electrical breakdowns were analyzed thread bare and found out the majority of the stoppages are due to `failure in Auto doffing functions` and OEM vendor had been called for retraining our Maintenance crew in both Mechanical and Electrical & Shift officers for trouble shooting as well as learning tips on maintaining the Auto doffers in Ring frames.
3. Traveller changing schedules were not followed in any proper order and trials were conducted for increasing the lifespan of travelers from the existing 8 days to 15 days (by proper selection of travelers) and merging one schedule traveler changes with Machine cleaning.
4. Lap Former breakdowns were analyzed for the root cause and it was identified as the pressure drop at infrequent intervals. This had been sorted out by keeping a storage tank nearby preparatory with standard pressure level so that Lap former machines are undisturbed.
5. Mills also planned not to stop the machines for any reasons except for breakdowns and on all Sundays mills standardized their Utilization levels to 99.5% plus.

With all these steps taken by the mills the team was able to achieve 98.9% within a period of 3 months and their efforts continued towards reaching the goal.

CS 17 Case study on "Yarn realization improvement"

In one Spinning mill, which was producing 30s and 40s Combed compact yarn for warp consumption, mills set the target of 74.5% as the standard yarn realization. Their standard of 74.5% was arrived at with the following process standards for the raw material based on its trash content and also based on Customer expectation:

Blow room waste level: 3%
Card waste level: 7.0%
Comber noil: 15%

In actual, Yarn realization in the mills for this process was found as 72.5% only. Upon analysis, it was identified that the waste at Cards alone has a deviation in 65% of the cards tested in that fortnight. Card waste level ranged between 7% to 9.5%. Hence, Zone-Wise wastes were collected at different critical zones and tested for lint content and trash content in all the Cards. The Quality assurance and the Maintenance team found the following deviations:

- Almost in all the cards investigated, the Pre carding zone waste level was higher by 0.5 to 0.8% and it has lint.
- Almost in all the cards the post carding zone waste level was higher than the std by 1.0 to 1.2% and predominantly contain lint.
- Out of 12 Cards investigated, only in three cards, Lickerin zone waste was more by 0.5% than the standards.

Zone-wise analysis of waste in cards had given a clear picture on lint loss in Cards and necessary setting corrections were made for achieving standards in all the zones separately. After doing the same, mills were able to achieve consistency in the waste level of 7% and the mills were able to achieve Standard Yarn realization of 74.5% therafter in this process.

Based on this experience mills revised their Carding waste analysis format by creating standards for each zone. Model format for zone-wise card waste analysis is enclosed in Annexure XV.

Annexure XII A: Model Spin Plan

	Unit 1					Unit 2		
Count	30s	40s	50	56	Total	80	Total	U1+U2 TOTAL
	CH	CW	CWCS	CHCS		CWCS		
Mixing	SFC	MS	MLS	MLS		ELS		
Type	Normal	Normal	Compact	Compact		Compact		
No. of RF-Proposed	5	9	2	6	22.0	20	20.0	42.0
Spindles	1200	1200	1200	1200		1200		50400.0
Wraping Count	30.0	40.5	50.5	56.5		81.0		
Spindle speed	19000	21500	22600	19350		23250		
TM	3.6	4.1	3.9	3.6		4.0		
TPI	19.4	25.8	27.7	26.7		36.0		
Grms/Spl	220	140	110	88		55		
P/Day @ 99 % Effy	3928	4495	787	1887	11097.9	3928.3	3928.3	15026
Waste %	3.0	3.0	1.8	1.8		1.5		
B.S.Req	4050	4634	801	1921	11406.1	3988	3988.1	15394
	117851	182054	39758	106624	446286.3	318194	318193.9	764480
Avg Count					40.2		81.0	50.9
SIMPLEX								
Speed(rpm)	1300	1300	1300	1300		1250		
Hank	0.85	0.85	1.20	1.20		1.60		

(Continued)

			Unit 1			Unit 2		
T P I	1.34	1.34	1.42	1.42		1.61		
T M	1.45	1.45	1.30	1.30		1.27		
Obtainable O.A.Effy %	82.00	84.00	85.00	85.00		88.00		
Prodn/Spl/8 Hrs	6.75	6.92	4.66	4.66		3.08		
Working Hrs	24.0	24.0	24.0	24.0		24.0		
Prodn/Spl/day	20.3	20.8	14.0	14.0		9.2		
No.ofSpls Req	200	223	57	138	618.0	431	431.4	1049.4
No.of M/Cs Req	0.91	1.01	0.26	0.63	2.8	1.96	2.0	4.8
M/C Allotted/Available	1	1		1	3	2	2	5.00
No. of RF/SMX	5.50	8.87	7.67	9.60		10.20		
No. of SMX/RF	0.18	0.11	0.13	0.10		0.10		
B.S.Req @ 0.5 % Waste	4070	4657	805	1930		4008		
FINISHER DRAWING	Twin delivery	Twin delivery	Twin delivery	Twin delivery		Twin delivery		
Speed (mpm)	450	450	450	450		400		
Hank	0.10	0.10	0.12	0.12		0.14		
Obtainable O.A.Effy %	88	88	88	88		90		
Prodn/2 Del./8 Hrs	2245	2245	1871	1871		1458		
Working hours	24.0	24.0	24.0	24.0		24.0		
Production/Day	6736	6736	5613	5613		4374		
No.of Deliveries Req	0.6	0.7	0.1	0.3	1.8	0.9	0.9	2.7
M/C Allotted/Available		2		1	3	1	1	4
No.of RFs/DF	8.3	13.0	13.9	17.4		21.8		
No. of DF/RF	0.12	0.08	0.07	0.06		0.05		
B.S.Req 0.1% W	4074	4662	806	1932		4012		

			Unit 1			Unit 2		
COMBER								
Speed (NPM)	550	550	500	500		400		
Hank	0.11	0.11	0.11	0.11		0.14		
Noils %	14.0	14.0	17.0	17.0		15.0		
Obtainable O.A.Effy %	88.0	88.0	89.0	89.0		90.0		
Feed length/nip in mm	5.6	5.6	5.2	5.2		4.8		
Lap wt (gm/met)	80	80	76	76		72		
100% Prodn/Comber/8Hr	653	653	511	511		370		
Working hours	24.0	24.0	24.0	24.0		24.0		
Prodn/Comber/Day	1958	1958	1533	1533		1110		
No.of M/Cs Req	2.08	2.38	0.53	1.26	6.2	3.61	3.6	9.9
Machines Alloted		5.00		2.00	7.0	4.00	4.0	11.0
No. of RF/Comber	2.40	3.78	3.80	4.76		5.54		
No. of Combers/RF	0.42	0.26	0.26	0.21		0.18		
B.S.Req Noils+0.2%W	4749	5434	974	2334	0	4731		18221.0
UNILAP								
Speed (mpm)	120	120	120	120		100		
Lap wt(gm/met)	80	80	76	76		72		
Obtainable O.A.Effy %	77.0	77.0	77.0	77.0		78.0		
Prodn./8 Hrs	3548	3548	3371	3371		2696		
Working hours	24.0	24.0	24.0	24.0		24.0		
Prodn/UL/Day	10644	10644	10112	10112		8087		
No. of M/Cs Req	0.45	0.51	0.10	0.23	1.3	0.59	0.6	1.9
M/C Allotted/Available					2		1	3.00
NO.of RFs/ LH10&UNILAP	11.21	17.63	20.77	26.00		34.18		

(Continued)

			Unit 1			Unit 2		
NO of UL/RF	0.09	0.06	0.05	0.04		0.03		
B.S.Req @ 0.2% Waste	4758	5445	976	2338		4741		18257.5
PRE COMBER DRAWING								
Machine Model		Super High Speed Twin delivery Drawframes(NAL)						
Speed(mpm)	700	700	700	700		550		
Hank	0.125	0.125	0.125	0.125		0.150		
Obtainable O.A.Effy %	85.0	85.0	85.0	85.0		88.0		
Prodn/DF/8 Hrs	2699	2699	2699	2699		1830		
Working hours	24.0	24.0	24.0	24.0		24.0		
Production per day	8097	8097	8097	8097		5489		
No of DF`s reqd	0.6	0.7	0.1	0.3	1.7	0.9	0.9	2.5
M/C Allotted/Available					2.0	0.86	1.0	9/12
No.of RF/PCDF	8.51	13.38	16.60	20.77		23.15		
No. of PCDF/RF	0.12	0.07	0.06	0.05		0.04		
B.S.Req 0.2% W	4768	5456	978	2343		4750		18294.1
CARDING		Single Lickerin Higher width Super High production Cards						
Speed (mpm)-OA	220	220	165	165		120		
Hank	0.100	0.100	0.120	0.120		0.140		
Obtainable O.A.Effy %	90	90	90	90		92		
100% Prodn/Card/Hr	70	70	44	44		28		
Prodn/Card/24Hrs	1514	1514	946	946		616		
No. of M/Cs Req	3.1	3.6	1.0	2.5	10.3	7.7	7.7	18.0
M/C Allotted/Available		7		4	11	8	8	19.0

(Continued)

		Unit 1			Unit 2	
No. of RF/Card	1.59	2.50	1.94	2.42	2.60	
NO. Of cards/Rf	0.63	0.40	0.52	0.41	0.39	
Waste %	7.50	7.50	7.50	7.50	8.50	
BS for Card	5154	5898	1057	2533	5192	19833.5
Waste% at Blow room	3.5	3.5	3.5	3.5	1.5	
Mixing required per day	5341.1	6111.8	1095.1	2625.0	5270.7	
Bales per day	31.42	25.98	6.40	15.35	21.63	100.8
(170 Kg Indian Bale/240 Kg-ELS bale)		57.4		21.8	21.6	100.8
AUTOCONER						
Production per day required at Autoconers with HW of 0.4%	3944.1	4513.2	790.4442	1894.722	3944.09-64	
Autoconer effective speed	1650	1650	1500	1500	1400	
Overall efficiency	76.0	76.0	78.0	78.0	82.0	
Production per drum per 24 hrs	35.6	26.3	19.7	17.6	12.1	
No of drums required	111	171	40	108	327	
No of Drums/Machine	72	72	72	72	72	
No of Autoconers Required	1.54	2.38	0.56	1.49	4.54	
No of Autoconers allotted				6	5	
Expected Yarn Realisation	74	74	72	72	75	

Annexure XII B: Production Formulae Used in Spin Plan

No	Department/Machine/ Factor	Formulae	Symbols used
1	Twist per inch	=Sq.root(count)*Twist multiplier	
2	Ring frame production in grams per spindle per 8 hours	=(7.2*S*PE)/(Wrapping count*tpi)	S=Spindle speed in rpm PE=Production efficiency
3	Speed frame Spindle point production in kg per 8 hours	=(7.2*S*OAE)/ (H*tpi*100000)	OAE=Overall efficiency H=Roving hank
4	Twin Delivery Drawframe production (Precomber DF and Finisher DF)in kg per 8 hrs	=2*(0.2835*DS*OAE)/ (H*100)	DS=Delivery speed in metres per minute H=Hank of sliver
5	Comber production in kg per 8 hours	={(0.0035*g*N*l)* (100-CN)*OAE}/10000	g=Lapweight in grams per metre N=Nipper speed in NPM l=Feed length in mm per nip CN=Comber noil in %
6	Unilap Production in Kg/8 hours	=(DS*g*480*OAE)/ (100*1000)	DS=Delivery speed in meters per minute g =Lapweight in grams per meter
7	Card production in kg per 8 hours	= (0.0354*DS*OAE)/ (H*100)	DS=Delivey speed in meters per minute H=Hank of sliver
8	Production per drum in 8 hours in Autoconers/ Linkconers	= (0.2835*ES*OAE)/ (C*100)	ES=Effective speed in meters per minute C=Count
9	Expected Yarn Realisation	=(Y/M)*100	Y=Yarn production in kg at Autoconers M=Mixing required for "Y"

Note: All these formulae are being used in designing the Spin plan in Annexure VIII A. Mills can design their own Spin plans using the excel file as a standard template model.

Annexure XIII: Model Mill Standard Document for Spindle Point Production in Grams /Spindle/8 Hours

No	Count & Description	Mixing	RF type	Wrap count	Average Spl speed	TM	TPI	Machine Efficiency	grams /8 hours/ spindle
1	**30s CHD**	**Indian**	**Normal**	**28.5**	**18650**	**3.55**	**18.95**	**96.5**	**240**
2	30s CHE	Indian	Normal	30	18750	3.6	19.72	96.5	220
3	30s CKE	Imported	Normal	30	19200	3.45	18.90	96.5	235
4	30s CHD	Indian	Compact	28.5	18650	3.4	18.15	95.5	248
5	30s CHE	Indian	Compact	30	18750	3.45	18.90	95.5	227
6	30s CKE	Imported	Compact	30	19200	3.3	18.07	95.5	243
7	30s CHD	Indian	Compact - Normal	28.5	18650	3.3	17.62	95.5	255
8	30s CHE	Indian	Compact - Normal	30	18750	3.35	18.35	95.5	234
9	30s CKE	Imported	Compact - Normal	30	19200	3.2	17.53	95.5	251
Note	**CHD**			**Combed Hosiery yarn for Domestic market**					
	CHE			**Combed Hosiery yarn for Export market**					
	CKE			**Combed yarns meant for Export Knitting**					

Annexure XIV: Model Chart for Estimation of Yarn Realization

No	Details for the Month of Jan 2021	
1	Opening stock	A
2	Cotton issue for Jan 2021	B
3	Closing stock	C
4	Cotton Consumption(X)	X=(A + B − C)
5	Opening stock of yarn	D
6	Yarns despatched	E
7	Closing stock of yarn	F
8	Yarn production	Y=(D + E − F)
9	Opening stock of Waste	G
10	Waste despatched	H
11	Closing stock of waste	I
12	Waste realized	W=(G + H − I)
13	Invisible loss	IL={X − (Y + W)}
14	Invisible Loss %	=(IL/X)*100
15	Yarn Realisation	=(Y/X)*100

Annexure XV: Model Chart for Zone-Wise Waste Analysis for Cards

Card No	Date	Mixing	Count	Lickerin	Pre cleaning segment	Flat strips (FS)	Post carding segment	Total
Standard		Mix 1	34s CH					
1								
2								
3								
4								
5			40s CH					
6								
7								
8								
Range								
Average								
Standard								

(*Continued*)

Card No	Date	Mixing	Count	Lickerin	Pre cleaning segment	Flat strips (FS)	Post carding segment	Total
9		Mix 2	80s CCW					
10								
11								
12								
13			60s CCW					
14								
15								
16								
Range								
Average								
Standard								

Note:
1. Total waste should be maintained at a level of plus or minus 0.5%
2. FS and Lickerin waste should be adjusted only when the actual waste is different from the standard by more than 15% even after rechecking.
3. Pre and Post cleaning segment waste needs to be adjusted only if these wastes are higher by 50% from their standards even after rechecking.

4

Training and Development

LEARNING GOALS

Development of Training and Development
PAT
Training standards and Periods
HPT-Human Performance Technology

Introduction

Training and development is an essential branch of successful management systems for sustaining the improvements and establishing continuous improvements. Training is not limited to Productivity alone, but it covers Safety at work and Personal motivation and Personal development. Organizations which are conscious of the role of training invest their focused time bound plans with a flexible budget. Money invested on training and development of employees will certainly pay back if one needs to deal with it commercially. Practitioners in Human resource, Organization development consultants, and Human resource managers realize that any investment on T&D activities should show a positive return and improve the bottom line. In the march of civilization, training is an essential ingredient from the Stone Age itself.

4.1 Definition of Training

As per Goldstein and Ford "Training is defined as the systematic acquisition of Skills, Rules, Concepts or attitudes that result in improved performance in another environment'. Hence training programmes are designed with a leadership of competent persons in the field in an environment similar to the actual work environment.

DOI: 10.1201/9781003257677-4

4.2 Chronology of the Development of T&D

Twentieth century saw dramatic changes in the world of work and in the twenty-first century appears to have continued the theme of change. Technological developments have revolutionized the work methods and many organizations expanded their operations globally. It opens up competition in their field of operations and the pace of change itself is speeded up. After the Industrial Revolution, mechanization was started which demanded formal training of operators. Initially it was confined to the Industry sector but highly developed techniques were applied in the fields of Agriculture, fishing, and forestry. During this period of Industrial revolution Taylor explained the theory of "Scientific Management'. He explained about the importance of training for,

- High productivity
- Low accident rate
- Low wastage
- Maximization of profit.

In the 1960s, OD-Organization Development gained acceptance and this was the most talked about techniques in Industrial training. By the 1980s, the Quality Circle concept gained importance and it calls for "Employee involvement in building the Quality" by developing circles within the industry and across the same kind of industries to promote healthy competition. By the beginning of 1990s, developments in the Information Technology field enlarged the ways and means of training and it is tailor made and systematic."

TABLE 4.1

Training and Development in the New Millennium

Background ideas on Organization/ Management	Employer led training activity
Learner focus	Career counseling
Skill development	Modern apprenticeships/Induction programme/ on the job training
E Learning	E learning groups(IT departments in industries)
Leadership	Organised workshops by Management consultants
Emotional intelligence	Managing diversity based team based training
Knowledge management	Project management and training

4.3 Scope of Training and Development

Scope of training has also been shifted from mere technical skills to interpersonal skills covering a wide spectrum ranging from manufacturing industries to service sector at micro level. Growing Quality consciousness among customers and Global competition is forcing organizations to have "Quality conscious trained and empowered employees." By "empowered employee" we mean that he is empowered to own the process in which he is involved and he should be able to stop the production if he feels the process is beyond controllable limits. It is aimed at laying the foundation of TPM in Training and Development.

4.3.1 Development of Employees by Human Resource Department

Human resource development is a process by which the employees are helped in a continuous and systematic manner to:

- Recruit the "right personnel" for the job needs through proper tools/analytical methods
- Provide Induction programme briefly for familiarizing the new recruits to the Organsational culture and also the total process
- Arrange orientation training programme in which he will be oriented to the specific "job role" and responsibilities
- With the help of established training standards and evaluation methods train the employees with the help of Trainers/Training officers. (Standards for Training periods and Evaluation measures are provided in Annexure XVI.)
- Train them on their critical work practices apart from training standards
- Acquire performance data once in year against training standards and work practices & re-sharpen their capabilities to improve their performance further
- Develop teamwork for achieving organizational goals

Part Analysis Training for shop floor employees

Part Analysis Training is a disciplined system of teaching the employees "skills" compared to conventional "unsystematic training'. This is a system developed by Coats Patrons International.

Advantages of PAT

It permits a greater degree of discipline in the teaching of skills

It enables workers to be trained to standard performance

It permits closer attention to the demands of the "Quality" during the impressionable training period itself.

It permits a detailed instruction on the origins of waste generation and possible steps to eliminate/reduce waste

It avoids the spread of the bad work practices associated with the "look and copy" training systems in conventional training.

It minimizes the training time

It minimizes the detrimental effect on output which operator training might have.

4.3.2 Mechanisms for Successful Training and Development System

To achieve the goals of the stated needs of the functions of Training and Development function in our mills, we need to install sub systems as mentioned below:

- Training records for every single employee from the level of Induction, Orientation, and subsequent follow up including critical work practices. (These records should be treated as similar to "Equipment History Records" in the organization.)
- Performance-linked wages schemes
- Potential appraisal and development with clear policies for rewards/Punishment from the stages of training itself
- Feedback and performance coaching
- Identification of the stems of the Shop floor in time (Best employees can be upgraded to be a mentor/trainer)
- Training to shoulder additional responsibilities as above
- Creation of Self-managed teams among the shop floor employees
- Organization Development programme for example Environment Management System implementation, 5S, TPM etc
- Rewards and recognition system
- Properly planned "Re-training" programme for all levels of employees

4.4 Management Development Programme

Role of today's Manager: Mintzberg (1975) said Managers serve three primary roles "Interpersonal, Informational, and Decision-making." To perform these roles effectively, their skills need to be developed. Katz and Kahn (1970) categorized the skill levels for Managers as:

- **Technical**
- **Human Skills**
- **Conceptual skills**

Human skills refer to the ability to communicate, motivate, and lead. Conceptual skills make it possible to consider the organization as a whole and evaluate the relationships which exist between various parts/functions. Such skills are concerned with the realm of ideas and creativity.

Effective Manager—How to become?
Peter F. Drucker suggested five habits of the mind to be an effective Manager:

- Management of time
- Result orientation
- Setting and keeping the priorities
- Decision-making
- Strength building

Training Wheel
The wheel of training starts circumferentially from Business needs, Identifying the training needs based on Business needs, Specifying the business needs, Translating the training needs into action, Planning the needs, and Evaluating the training to achieve desired results. The centre is occupied by the people or the employees of the organization.

Training is an organized activity to impart information or instructions to improve the trainee's performance or to help him in realizing his required level of skills/knowledge.

4.4.1 Objectives of the Management Development Programme

Experts in HR development have drawn different objectives based on their experience and expectations:

Most essential objectives identified are:

1. To improve the performance of them in their current assignment
2. To provide adequate cover to the organization in the event of deaths or transfers
3. To raise the general level of Management thinking
4. To meet the anticipated needs of the organization
5. To improve cross functional relationships
6. To improve analytical ability
7. To understand the problems of human relations
8. To stimulate creative thinking

Role of Learning in Management Development

Learning of knowledge and skills through training can be enhanced by various factors such as Motivation, Response and reinforcement, feedback, participation and knowledge and perception of trainees.

Learning can be described as a relatively permanent change in behavior that occurs as a result of insight, practice or experience. Learning can be an addition or subtraction(unlearning a bad habit) or it could be a modification(changing from old to new techniques).Learning may be conscious or unconscious. For example learning the "acting in movies" is consciously done but the style and gestures is unconsciously acquired through films we have seen.

To motivate us to learn, we need different strategies. Stronger the motivation faster and effective will be the learning process.

Training Deliverables

Training deliverables can be defined as "end product of instructional design process." For example, workbooks/manuals/lesson plans/takeaways of the programme/training content in DVD etc are few to be classified into deliverables.

Training Techniques

Choice of training techniques is decided by:

- Learning objectives
- Size of the target population
- Learning styles and interest of the trainees
- Course contents/duration

Normally we classify these training techniques broadly into "On the job training and Off the job training." On the job training is most preferred as it is faster than off the job training. However sometimes we may need to use the technique of Off the job training too, like class room coaching with samples/defects/customer samples, field trial reports, video shows of right work practices, Industrial engineering case studies, Role playing techniques, Group discussions etc.

TABLE 4.2

Training and HRD Process Model

Assessment	Design	Implementation	Evaluation
Assess needs	Define objectives		Select evaluation criteria
	Develop lesson plan	Deliver the HRD program or Intervention	Determine evaluation design
Prioritize needs	Develop/acquire materials		**Conduct evaluation program or intervention**
	Select trainer		
	Select methods/techniques		
	Schedule the program/intervention		Interpret results

Assessment of Training Needs

Managers get the work done through coordination and direction of the efforts of others. Managers are organization members responsible for planning, organizing, leading and controlling the activities of the organization so that its goals can be achieved. For the business to improve, the skills of its managers need to be improved. Hence, managing for success requires a comprehensive set of managerial skills.

Need analysis involves investigating how the training could solve the issues related to performance or enhancing the current level of performance to a higher platform. Need assessment is the process of determining KSA's. (Knowledge/Skill/Attitude). This helps organization to plan their futuristic training requirements for their Managerial executives and also to allocate funds towards the same.

4.5 HPT-Human Performance Technology

If "needs" are to be unearthed then the performance level has to be measured. Hence HPT-Human Performance Technology is introduced. HPT is defined as a set of methods and processes for solving problems or realizing the opportunities related to the performance of the people. This can be utilized to discover the important human performance gaps, planning for future improvements and design & development of training programmes.

HPT covers five major areas:

1. Performance Analysis
2. Cause Analysis
3. Intervention Analysis
4. Change Management
5. Evaluation

Five Habits of the Mind to become an Effective Manager

Peter F.Drucker has outlined five habits of the mind to become an effective Manager which are more relevant in this context.

A. *Management of Time*
Time management is a tool in the hands of a Managerial executive to enhance his effectiveness in his Performance. Paradox of time must be understood and we should prioritise our activities so as to achieve the goals of the organization. This may cover daily management of time/eliminate non value added activities/controlling paper work/checking balance(similar to cash management) etc. Simply, we should learn to spend more time on activities which contribute to performance level enhancement.

B. *Result Orientation*
We need to consistently deliver required business results, sets and achieves achievable, consistently complies with Quality, productivity, timely deliveries and customer satisfaction etc. This calls for goal setting and drives the team for meeting/ exceeding expectations.

C. *Setting up the Priorities*
Setting up the priorities of the Managerial/professional position is an important element of an effective manager. Managers

should be trained to set goals and specifically should spend time on productive tasks that brings value to their contribution.

D. ***Decision-Making***

Decision-making skills are important for the managers to be effective in their performance and also to meet the requirements of the goals of the organization. Persons around him look for his decisions day in and day out and he should have the skill of identifying the real issues and try to help them resolve on every single occasion. Decision-making process involves making a choice from two or more alternatives/selection of alternative solution/Implementation and evaluation.

E. ***Strength Building***

To be a highly successful Manager, he should have the awareness of the strengths of each team member and able to discover their hidden talents which could be tapped for organization's goals. It involves the following steps:

- Identify the core strength(core competency level)
- Apply these core competency to organizations` goals
- Assign mini projects to them based on employee strengths
- Incorporate core competency levels in performance reviews
- Communicate team members on their career growth based on strength improvement
- Provide training facilities in to enhance their core strength
- Allow the team members to take higher responsibilities

CS 18: Case study on "Improvement in Work Assignment in Ring Frame through Re-training"

In one of the Spinning mills with an age of 15 years, there were many trained employees and mills intended to revise their work assignment. Existing status of work assignment in Ring frames for the tenters were as below:

1. No of spindles per Ring frame - 1200
2. Total number of Ring frames - 24
3. Ratio of Relievers against tenters - 25%
4. Average count - 40s Combed warp
5. End breaks level - 5-6 breaks per 100 spindle hours

6. Pneumafil waste - 2.8%
7. Number of Spindles looked after by one tenter - 2400
8. Hence the number of tenters per day - 3*12 i.e. 36 tenters per day
9. Total number of relievers per day - 3*12*0.25 i.e. 9 relievers per day
10. Total number of employees in Ring frame per day = 45
11. Employee wages per day(all inclusive) = Rs 550/-

Industrial engineer had studied the tenter"s actual piecing rate per hour separately by covering all the tenters. During the initial training provided to all these employees, all these employees were trained to piece minimum 12 piecings per hour and in 2400 spindles they pieced 280 piecings per hour. However as per the study conducted at the mills, the average piecing rate per hour was found to be 180-200 only. This is due to the following important reasons:

1. Upgradation of machinery has resulted in improved working performance
2. Mills has improved the raw material quality and this has resulted in end breaks rate significantly from 6 to 8 breaks per 100 spindle hours to the current level of 5 to 6.
3. However, due to poor follow up by Training and Industrial engineers towards re-training of employees resulted in lethargic work performance with the lower piecing rate.

Hence Mills decided to retrain the employees in batches and their piecing rate was improved to their original performance within a week`s time. As their piecing efficiency had been improved, the Management had revised their work assignment from 2400 spindles to 3000 spindles after discussion with the employees.

Employees also had agreed for a reduction on the requirement of relievers from 25% to 20% also. Now the financial impact of the retraining was estimated as follows:

1. Total number of spindles = 24*1200 i.e. 28800
2. No of spindles per tenter = 3000
3. No of tenters required per Shift = 28800/3000 = 9.6 i.e. 10 Tenters per day

4. No of tenters per day = 3*10 i.e 30
5. Relievers per day = 30*0.2 i.e 6 per day
6. Total number of Tenters and Relievers per day = 36
7. Difference in employee requirement in Ring frame before and after retraining = (45-36) i.e. 9 per day
8. Net savings per day = 9* Rs 550 i.e. Rs 4950 per day
9. Net savings per year = Rs 17.82 lakhs

Conclusion: Mills was able to save Rs 17.82 lakhs though retraining in Ring frame department for Ring frame tenters alone.

CS 19: Case Study on "Improvement in Work Assignment in Autoconer Department through Retraining"

In one of the Spinning mills with an age of 8 years, there were many trained employees and mills intended to revise their work assignment. Existing status of work assignment in Autoconers for the tenters were as below:

1. No of drums per Autoconer = 60
2. Total number of Autoconers = 8
3. Number of relievers = One reliever for every three tenter
4. Average count = 40s Combed warp
5. Clearer cuts level = 45 breaks per lakh meters
6. Number of drums looked after by one tenter = 48
7. Hence the number of tenters per day = 3*(8*60/48) i.e. 30
8. Total number of relievers per day = 30/3 i.e 10 per day
9. Total number of employees per day = 40
10. Employee wages per day(all inclusive) = Rs 450/-

Industrial engineer had studied the tenter`s actual Cop replenishing rate per minute ad per hour separately by covering all the tenters. During the initial training provided to all these employees, all these employees were trained to replenish 35 cops per minute with 1100 cops per hour on an average. However as per the study conducted by Industrial engineer, the average replenishment rate was found to be 26 cops per minute and 750 cops per hour only. The shop floor production team joined with Industrial engineer and training officer to investigate the root cause for such a deviation in the level of performance by the Winding tenters and assigned it as below:

A. Unnecessary clearer cuts were reduced in Autoconer due to improved maintenance practices in Spinning and preparatory.
B. Mills had improved the raw material quality and this has resulted in reduction of clearer cuts from 65-75 level during original work assignment study earlier and over a period it had been reduced to 45 level due to improved raw material and process conditions.
C. However, due to poor follow up by Training and Industrial engineers towards re-training of employees, employees had a free time which resulted in lethargic approach leading to poor work efficiency.

Hence Mills decided to retrain the employees in batches and their cop replenishment rate to the original level of 35 cops per minute and above 1100 cops per hour. After thorough retraining, all the Autoconer tenters reached the target level of performance. As the cuts were maintained at 45 level only, Management discussed with the employees on the following:

- Tenters would look after 60 drums instead of 48 drums
- Relievers would be provided at the rate of one per every four Tenters instead of one for every three.

Now the financial impact of the retraining was estimated as follows:

A. Total number of drums = 8*60 i.e. 480
B. No of drums per tenter = 60
C. No of tenters required per day = 3*(480/60)
   = 24 Tenters per day
D. Relievers per day = 24/4 i.e. 6 per day
E. Total number of tenters and Relievers per day = 30

F. Difference in employee requirement in Autoconers before and after retraining = (40–30) i.e. 10 per day

G. Net savings per day = 10* Rs 450 i.e. Rs 4500 per day

H. Net savings per year = Rs 16.2 lakhs

Conclusion: Mills was able to save Rs 16.2 lakhs per year though retraining in Autoconer department for Tenters alone.

Annexure XVI: Standards for Training Period and Evaluation Measures

Category of operative	Standards	Training period	First evaluation	2nd evaluation
Carding tenter	3 sliver piecing per minute	15 days	15 days	30 days
Pre comber Draw frame	3 sliver piecing per minute	15 days	15 days	30 days
Unilap Tenter		15 days	15 days	30 days
Comber tenter	6 lap changes per minute	15 days	15 days	30 days
Finisher Draw frame	3 sliver piecings per minute	15 days	15 days	30 days
Speed frame tenter	4 roving piecings per minute	15 days	15 days	30 days
Speed frame Doffer	20 bobbin doffs per minute	30 days	30 days	45 days
Ring frame tenter	12 ends per minute piecing and 280 ends per minute	30 days	30 days	45 days
Ring frame doffer	120 spindle doffs per hour and 16 ends gaiting per minute	30 days	30 days	45 days
Autoconer tenter	36 cops per minute and 1000 cops per hour in a Magazine Autoconer	30 days	15 days	30 days

5

Energy Management

LEARNING GOALS

- **Need of Energy Savings in a Spinning Mill**
- **Power Consumption Standards**
- **Energy Audits**
- **Energy Conservation Measures**
- **Harmonics and Harmonic Filters**
- **Energy Efficient Motors**
- **Rewinding of Motors**
- **MD Controller**
- **Advances in Motor Technology**
- **Renewable Energy**

Introduction

Energy cost occupies a major chunk of yarn manufacturing cost next to Raw material cost in spinning mills. Hence to reduce the manufacturing cost, it is essential to track the energy consumption in every segment of the Manufacturing operations and take innovative steps to control and bring it to the lowest level possible and also take measures to benchmark our mills against the industry standards. As a comparison, power tariff across the globe against India is given as follows:

An increase of Re 0.10 per unit will cost Rs 28.8 lakhs per annum for a mill having 50000 spindles with an average count of 50s. Within India itself, mills in Gujarat are realizing the best advantage of lower power tariff and Maharashtra bears the brunt of higher tariff. Other states in India are charging anywhere between Rs 5 to 6 per unit. Hence it is essential for the Spinning mills to save energy to reduce the Manufacturing cost and improve profitability.

DOI: 10.1201/9781003257677-5

TABLE 5.1

Power Tariff Per Unit-International

Country	HT tariff per unit in Rupees
Japan	15.5
China	6.40
Vietnam	5.75
Indonesia	5.5
USA	5.02
India*	4.25 to 7
Note*	Highest in Maharashtra and the lowest in Gujarat state

5.1 Power Consumption in a Spinning Mill

As per the Industry experience, the power cost in Spinning mills occupy the second highest percentage next to Raw material and it ranges from 10% to 18% depending on the products. However mills have several means of bringing it down by investments on Wind mills/Solar power installations/Open access trading etc. All these are not an assured return and it also requires capital investments except open access trading. In the case of open access trading it is purely dependent on the "supply-demand" situation and there is no guarantee for procurement as well as the price per unit. Hence it is definitely necessary for the Spinning mills to control energy consumption through suitable measures by dedicated involvement of the team.

Every mill must create a standard power consumption chart for the Spin plan for that month based on the "overall efficiency level" considered in each department. (A Model chart for Standard Power Consumption in a Spinning mill with 50000 spindles is enclosed for reference in Annexure XVII). Once it is made ready it is the responsibility of the Electrical team to monitor shift wise consumption and act on the deviations noticed every day. Deviations in power consumption are due to variety of reasons and major deviations are given below:

- Machines underutilized due to RM shortage/Breakdowns/Manpower shortage
- Absence of lubrication in few machines(less spindle oil in spindle bolsters/Absence of greasing in drive shaft bearings in Ring Frame, worn belts in drive pulleys/rewound motors/Department temperature)
- Malfunctioning of Auto controls of Humidification plants

- Waste clogging in Filters of Automatic waste evacuation system and Humidification plants
- High incoming voltage
- Higher level of harmonics
- Part utilization of any machine(One side of RF only running/ only one delivery of comber is running etc)
- High level of end breaks and pneumafil waste in Ring frames
- Higher level of clearer cuts in Autoconers
- Higher level of sliver breaks in Cards (Blow room as well as cards are kept running while the sliver is broken)
- Higher system pressure in Air line
- Air leakage in critical machines like Autoconers, Lap formers etc

The list doesn't end with these points as there could be many more reasons and shop floor executives along with the Electrical team need to identify the exact reason and eliminate its impact altogether. The analysis should be treated similar to production deviation reports by the Top Management in their periodical reviews.

5.2 Power Consumption–Industry Norms for Spinning Mills

Mills have the practice of estimating Units per kg consumption and SITRA has published conversion factors for "40s UKg" similar to grams per spindle. Apart from the same, industry now refers to "Unit per count per Kg" for benchmarking their unit's performance on power consumption, a newer but more useful parameter for shop floor controls.

TABLE 5.2

Industry Norms-Power Consumption

No	Counts	Very good	Good	Poor
1	24s to 49s	0.095	0.105	0.115
2	>50s to 99s	0.105	0.110	0.120
3	>100s to 120s	0.115	0.125	0.135

If we estimate the impact of Losses on account of energy for a 50000 spindle mill with 50s Average count between "Very good" mills and "Poor" rated mills, it amounts to a significant loss per month as explained below:

TABLE 5.3

Impact of Higher Unit Per Count Per kg for a 50000 Spindles Mill

Factors	Rating: Very good(0.105)	Rating: Poor(0.120)
No of Spindles	50000	50000
Average count	50s Combed Compact Warp	50s Combed compact warp
Spindle point production in grams /8 hrs	106	100
Production per day	15900	15000
Units consumption per day	83475	90000
Difference in consumption per day		6525
Cost per unit in Rupees		5.90
Loss per day in Rs		38497.5
Loss per month (Rs in Lakhs)		**11.55**

Hence the mills must be aware of the draining of income by way of energy losses due to excess power consumption.

5.3 Innovative Projects for Energy Conservation

Mills should initiate innovative projects apart from regular energy monitoring systems as below:(listed in an order of priority)

1. **Up-gradation of Machines with energy saving concepts:**
   - Centralized waste collection
   - Centralized compact suction
   - VFD to replace variator drives
   - Inverter controlled suction motors in Autoconers
   - Inverter controlled Pneumafil motors
   - Energy efficient Ring tubes(Slim tubes)
   - Lower Lift and Ring diameter or optimization of Lift and Ring diameter etc

2. **Study of Humidification plants and Automatic waste evacuation systems for energy losses and take suitable actions like:**
   - Inverter drives for pumps and fans
   - Design of filter
   - Automatic cleaning of filters with a timer
   - Advanced design of eliminators and Nozzles etc
3. **Replacement of all Motors having more than 4 rewinding*** (Due to its importance rewinding of motors is separately covered too) and replace these with Super energy efficient IE3/ IE4 motors
4. **Air leakage audit** needs to be conducted and Air consumption and Power consumption for Compressor must be brought to standard level of Performance. Ideally power consumption for a compressor should not exceed 3.5 % of the total unit consumption in a medium scale modern mill today.
5. **Transformer loads should be redistributed** so that every transformer is loaded in a range of 50-60% load to maximize its efficiency. This must be done at the time of Planning the projects itself and there should be a flexibility to accommodate further transformers in case if the mills had plans to expand Spindles in the same unit.
6. **Install "On load tap changer**"for incoming main Transformer so as to control the incoming voltage. Control on incoming voltage is an essential factor in controlling the power consumption. Not only power consumption, but also it helps to eliminate erratic function of Clearers/electronic devices at higher levels of voltage. It also helps to control power factor as at higher voltage power factor decreases significantly.
7. **Install Automatic power factor Controller** and ensure the adequacy of capacitor bank to improve power factor.
8. **Use VFD's** (Variable frequency drives) wherever it will be beneficial to save energy like blowing fans, suction fans, intermittent stop and start machines/equipments etc.
9. **Use Energy meters** in vulnerable areas and monitor the machine wise /section wise consumption/analyze the deviation/ take actions immediately. Suggested critical machines are:
   - Autoconers
   - Ring frames
   - Humidification plants

- Compressors
- Automatic waste evacuation plants (In all the other machines it is enough to measure the energy from respective SSB itself and it is not necessary to estimate machine wise)

10. **Install adequate and appropriate Harmonic filters** so as to control the ill effects of harmonics on energy consumption
11. **Install Maximum Demand controller** to optimize the Maximum demand. By installing MD Controllers mills can eliminate the impact of instant loads on energy cost.
12. **Audit the Mill's Lubrication programme** and take necessary actions, as improperly lubricated parts tend to increase energy consumption. This audit help the mills to:
    - Revise the schedule of lubrication in all the machines, especially,critical machines like Ring frames
    - Change the type of lubricant which retains its characteristics for a longer duration
    - Utilize the advanced technology of Lubrication towards energy consumption
13. Wherever the machines are running at less load of 40% intermittently, **Automatic Star-Delta-Star converters** can be installed.
14. **Use of either flat belts or cogged belts** instead of conventional V belts to save minimum 3 % on energy in that drive.
15. Use of LED bulbs in factory and environment lighting to save energy on lighting needs considerably

5.4 Measures to Reduce Energy Consumption in Compressors

- Energy efficiency of the compressor must be assessed while procuring itself by User reviews. Today fourth generation Compressors are available with a specific power consumption as low as 0.14 unit per cfm (First generation Compressors are with 0.24 and third generation compressors are offered at 0.16)
- At the time of installation of Compressors proper ventilation for the Compressor room must be ensured. Hot air from the compressor must be let out at a sufficient height from the Compressor room.

- Standard pipe line dimensions/flow meters/standard tubes and fittings as per manufacturer's recommendations should be followed. Any violation in these specifications will lead to probable higher air consumption or air leakage in the subsequent period.
- Over head Air line with a closed loop must be the choice as it will help to detect leakages immediately
- Refrigerated driers should be preferred as it is energy efficient and maintenance of dew point is quite easier with proper radiation.
- At necessary places, storage tanks need to be provided to maintain pressure level required for critical machines like Ring frames/Lap formers/Autoconers.
- Auto draining of collected water from the airlines must be incorporated at Storage tanks /driers etc.
- It is advised to provide a separate Compressor airline for cleaning activities or Mobile compressor can be preferred for departments where air is used for cleaning.

Mills must establish a standard procedure for air leakage measurement periodically and control. Operations on air leakage involves:

- Estimation of standard air consumption requirements for the mills
- Setting the system pressure at 7 kg/sq.cm
- Maintaining the driers properly with a dew point temperature below 4 degree centigrade
- Provision of auto draining in driers and storage tanks
- Monitoring the temperature in the compressor room
- Maintaining quality of the components used in Pneumatic connections including tubes/fittings/solenoids/valves etc and maintaining the same
- Periodical Air audits for assessing the Air leakage/Devoted air leakage checking on Holidays
- Estimation of Compressor loading and Unloading once in a month without fail to assess the extent of Air leakage
- Tracking the power consumption by Compressor on a daily basis
- Thermo-graphic analysis of Compressor motor and to take necessary actions immediately

5.5 Power Quality-Harmonics

Quality of power is more important which decides the efficiency of the motors and other electrical devices including power electronics components. One of the critical factors affecting power quality is "Harmonics." Harmonics are voltage and current frequencies riding on the top of the normal sinusoidal wave forms. (In a common man's language this is nothing but a disturbance to both voltage and current). These Harmonics are created in multiples of the fundamental frequencies i.e. 50 Hertz. Harmonics are normally created by non linear loads(switching loads). The higher frequency waveforms are normally called "Total Harmonic Distortion-THD," which perform no useful work but it can be a significant nuisance.

5.5.1 Sources of Harmonics

Harmonics are created by the presence of:

- Static power convertors/Rectifiers as used in UPS/Battery Charger
- Power electronics used for motor controls/Invertors
- Saturated Transformers
- Fluorescent lighting

5.5.2 Effects of Harmonics

Harmonics offers significant nuisance for any electrical system and its major ill effects are:

- Causes additional losses in Motors/Transformers/similar electrical devices
- Heat development in Conductors
- Nuisance tripping of relays and capacitors
- Leads to lower power factor and increases electrical bill
- Due to the heat development, insulation in electrical equipment will be spoilt

5.5.3 Selection of Harmonic Filters

Harmonic filters need to be selected and incorporated in our Electrical distribution system at suitable locations considering the following factors:

- KVA requirements of the load
- Harmonic profile of the load current
- Harmonic factor of the Neutral current
- Configuration of the existing or proposed electrical system

Mills can choose on the type of the filters from the following types by considering the above criteria:

- Passive Harmonic filters: Provide low impedance path to Harmonics and ground or Create high impedance to discourage the flow of Harmonics.Passive filters cannot adapt to changes in the system
- Isolation transformers: These are filtering devices to segregate the Harmonics from the circuit in which they are created, protecting upstream equipment/devices from the ill effects of Harmonics.
- Active Harmonic Filters: These filters continuously adjust their behavior in response to the harmonic current content of the monitored circuit.These are designed to accommodate a full range of operating conditions without any further intervention.

5.6 Energy Audit

Energy Audit with the help of an external expert in energy conservation programme in Spinning mills can be made to conduct Energy Audit at least once in 3 years period to:

- Identify the scope of further energy reduction
- Identify the obsolescence in Systems and practices
- Implement advanced technology in our systems i.e. similar to centrifugal fans for Humidification/Super premium energy efficient motors to replace 4 times rewound motors etc
- Assess the shop floor benefits realized from Prediction equipments(PDM tools) like Air leakage detector/Load manager/ Energy meters/Thermography analyser/Ultrasonic detectors etc
- Assess the efficiency of the existing Energy Management programme in the mills and to arrive at an action plan to plug the drain, if identified.
- Assess the Energy efficiency level of Compressor/Drier/Air line system

Based on the Audit report an action plan needs to be submitted by the Head of Electrical(Engineering) to the Top Management with defined activities and target dates for completion towards improvement actions.

5.7 Rewinding of Motors

Mills has several practices of rewinding of motors as below:

- Replacement of motors after four rewinding
- Replacement of motors only in case of burnouts
- No history of rewinding

But the Energy conscious mills follow the following practice:

- Rewinding history of motors is tracked by painting in the base plate itself(Green-1st Rewinding/Yellow-2nd Rewinding/Red 3rd Rewinding). After third rewinding mills run this motor and when there is a need for fourth rewinding it should be disposed off.
- However, rewinding of motor needs a proper tracking on the quality of rewinding and Motors after rewinding must be tested as per the procedure explained in Annexure XVIII for "Test procedure for assessment of Rewound motors."

5.8 Energy Efficient Motors

As per IEC(International Electro technical Commission) standards, the energy efficiency of motors are continuously improving and the energy efficient motors are ranked as below:

IE1 - Standard Efficiency
IE2 - High Efficiency
IE3 - Premium Efficiency
IE4 - Super premium Efficiency

As per the standards provided, the Efficiency of 4 pole motors for the normal range of 20 KW to 65 KW in Spinning mills, the efficiency difference between IE1 and IE4 will be 5 to 6% and between IE1 and IE3 it will be 3-4%.

Mills need to constantly evaluate the power consumption in motors (which depends on the characteristics of the load and condition of the motors) and take appropriate actions to restore the optimum/best performance conditions. The average life of the Induction motors is given below:

TABLE 5.4

Average life of Induction Motors

No	Motor capacity	Expected life in years
1	Less than 7.5 KW	12
2	7.5 KW –250 KW	15
3	Above 250 KW	20

Even though the life is specified as above, mills are running above these limits because of their ambient conditions provided to motors and also the Maintenance. However over this period, quality of winding and technology has improved with the onset of Energy efficient motors. These energy efficient motors can be used to replace old motors where:

- Existing motors have crossed rewinding norms
- Application of these motors is for utilization efficiency above 18 hours
- Load on the motors is above 75%
- No of "stops and starts" are lower
- Requirement of energy efficiency is essential
- We plan to convert the drives with VFD/Inverters

As per the statistics provided by BEE(Bureau of Energy Efficiency), average payback for replacing an IE1 motor by IE3 motor is less than 2 years.

When we invest in motors, we need not give more weightage to the price of the motors as its price forms only a small portion of the life cycle cost. The life cycle cost of a motor can be illustrated by the following example:

1. Capacity of the motor = 45 KW
2. Utilization efficiency = 85%
3. Motor efficiency = 92%
4. Energy consumption in units per hour = $45 * 0.85 * 0.92$ = 35.19 units per hour
5. Energy cost per life of 15 years = Rs 277 lakhs

6. Average annualized maintenance cost for the motor for one year = Rs 7500
7. Maintenance cost per 15 years = Rs 112500/-
8. Cost of the 45 KW motor = Rs 1,40,000
9. Total life cycle cost in lakhs = Rs(277 + 1.125 + 1.4) = Rs 279.525
10. % of Motor cost to its life cycle cost = (1.4/279.525) $*$ 100 = 0.5%

This illustration makes it clear that we need to work on energy efficiency of the Motor rather than the "purchase price of the motor," which is negligible. When the motor is inefficient due to more rewinding or if we choose a sub-standard energy efficient motor due to low cost-we will end up with higher "operating cost of the motor," which adds to our energy cost.

5.9 Renewable Energy

Non-renewable, or "dirty" energy includes fossil fuels such as oil, gas, and coal which results in carbon emissions and other forms of pollution. Towards a clean environment globally research is going on for effective use of renewable energy sources. This move also helps to reduce the dependence of any country on oil rich nations for fossil fuels.

Renewable power generation is in increasing trend with a global vision of "Net Zero by 2050." Let us have a look at global and Indian scenario on Wind energy and solar energy, which stands next to Hydropower as second and third respectively.

TABLE 5.5

Global Renewable Energy Status as on 2021

Type of Renewable energy	India	World
Wind generation	39 GW	733 GW
Solar power generation	40.1 GW	714 GW
Hydropower	46 GW	1211 GW

As wind power helps to reduce power cost in Spinning mills, many mills have invested to the extent of 100% of their consumption to save energy cost. As a move towards fulfilling the Global vision 2050, Solar energy is also being promoted globally across all nations and investments on solar power generation plants by the Spinning mills is happening in India following the same pace of other nations.

CS 20: Case study on "Incorporation of MD controller"

In one Spinning mill with 50000 spindles, the Sanctioned demand was 4000 KVA. As per the EB tariff procedure, MD charges were being estimated and billed on "maximum demand reached" in that billing period. Upon analysis of the MD charges paid by the mills for a preceding period of 6 months, it is identified that:

- Peak demand reached in all the 6 months were 3650,3700, 3665,3690,3670 and 3750 KVA
- The corresponding next highest level of Demand reached in the same 6 months period were only 3450,3500,3490,3525,3480 and 3520
- Hence it is identified that there is a minimum scope of 125 KVA which can be controlled.

Hence mills invested on MD controller and fitted in the line with safety precautions and the Limit is set at "3550 KVA" on a safer side. The MD controller tripping mechanism is connected to SSB which is controlling one Humidification plant to switch it off, whenever there is a rise in demand crossing the set value of 3550 KVA. After implementation, in the first month the mills savings on MD charges itself is Rs 30000/-(@Rs 300 per KVA).

Cost of installation of MD controller = Rs 40000/-

Hence the amount invested on MD controller is paid back within 40 days and the rest is recurring savings on energy cost.

CS 21: Case study on "Energy Savings in AWES-Automatic Waste Evacuation System"

In one Spinning mill, the Centrifugal fan in the Automatic waste evacuation system was running with V pulleys (4 numbers of B groove). During the Energy Audit by the internal team, it is decided to opt for a Flat pulley system with inverter control. Existing conditions were assessed as a controlled study which is given below:

1. Motor capacity = 55 KW
2. Speed = 1400 rpm
3. Unit consumption = 47 units per hour

Flat pulleys (both driver and driven) were incorporated and flat belt drive is mounted for the Centrifugal fan. After adjusting the tension control for the drive system, load study is repeated for the centrifugal fan. Inverter is also connected to maintain standard suction level uniformly. As per the load study power consumption with flat belt and inverter was estimated as 38 units per hour only. It means the savings per hour is 9 units per hour which translates into 19% on energy consumption in AWES.

Based on this study mills decided to convert drive systems in one more AWES In the same mills and achieved the same level of savings in that AWES also.

CS 22: Case study on "Inverter control for Pneumafil in Ring frame"

One Spinning mills decided to study the usefulness of Inverter controls for energy savings in their Shed. Details of the Ring frames are given below:

1. Total number of Ring frames : 24
2. Number of spindles per Ring frame : 1008
3. Pneumafil motor capacity : 6.5 KW

4. Count spun in the shed : 60s & 80s Combed compact weaving
5. End Breaks maintained (average) : 4 per 100 spindle hours
6. Pneumafil waste level : 1.8% and less

It is decided to conduct an energy study before and after the installation of inverter controls for the Ring frame. Hence one Ring frame running on 60s Combed compact weaving yarn is taken for controlled study.

As per the study, energy consumed per hour is estimated as 4.4 units per hour based on two full doff studies so as to have a realistic estimate. Mills procured Inverter and its controls from OEM supplier of the Ring Frame itself and repeated the controlled study. In the same manner mills had conducted two load studies for two full doffs and found out that the energy consumption is reduced to 3.2 units per hour. This helps the mills to realize 1.2 units per hour in one RF.

1. Savings in unit/hour = 1.2
2. Savings per day = 1.2 * 24 i.e. 28.8 units per day
3. Unit cost in rupees = Rs 6.5
4. Savings per month = Rs 6.5 * 28.8 * 30 i.e. Rs 5616
5. Cost of Inverter and its controls from OEM vendor = Rs 30000/-
6. Payback period in months = (30000/5616) months i.e. 5.3 months

A payback period of 5.3 months is highly attractive and it helps to save energy in Ring Frame pneumafil by more than 25%.

Based on this result, mills decided to provide Inverter controls in all the Ring frames in the shed to save energy further.

Annexure XVII: Model chart for Standard power consumption in a Spinning mill with 50400 spindles.

Power Consumption Departmentwise Standards in a Spinning Mill-50400 Spindles

Department	Name of M/c	Number of M/C	KW/ M/C	Ut %	Load %	Total	Cons/ day	% of Total power
BLOW ROOM	Blendomat	1	12	0.8	0.4	3.84	92.16	0.11
	Uniclean	1	10	0.8	0.4	3.20	76.80	0.09
	Multimix	2	12	0.8	0.4	7.68	184.32	0.22
	Flexiclean	2	7.5	0.8	0.7	8.40	201.60	0.24
	Suction fans and Exhaust fans	4	4	0.8	0.7	8.96	215.04	0.26
Cards	TC 5-1	24	17	0.9	0.6	220.32	5287.68	6.42
Unilap		3	12	0.85	0.7	21.42	514.08	0.62
PCD	SB20	3	12	0.85	0.5	15.30	367.20	0.45
LK 64 Z Combers		12	12	0.9	0.4	51.84	1244.16	1.51
AWES for cards and combers		2	80	0.9	0.85	122.40	2937.60	3.57
Rieter D22 DF		4	12	0.9	0.5	21.60	518.40	0.63
Speedframes	LF 4200	12	27	0.8	0.5	129.60	3110.40	3.78
Spinning	LR 60A	50	45.5	0.99	0.75	1689.19	40540.50	49.21
	OHTC	50	3	0.99	0.7	103.95	2494.80	3.03
	Suessen Suction	50	7.5	0.99	0.85	315.56	7573.50	9.19
Autoconer	Autoconers	16	18	0.82	0.75	177.12	4250.88	5.16
H.PLANTS-SUPPLY		10	22.5	0.95	0.8	171.00	4104.00	4.98
EXHAUST		10	18.5	0.95	0.8	140.60	3374.40	4.10
PUMP		10	7.5	0.6	0.9	40.50	972.00	1.18
COMPRESOR	Ingersol Rand	1	110	0.95	0.9	94.05	2257.20	2.74
	Claening Compressors	2	45	0.2	0.85	15.30	367.20	0.45
	Driers	3	5	0.45	0.7	4.73	113.40	0.14
LIGHTING		1	110	0.99	0.6	66.00	1584.00	1.92
Total unit consumption per day							**82381.32**	

Annexure XVIII: Test Procedure for Assessment of Rewound Motors

Gate pass no
Date

Name plate information

Make	Year	KW	rpm	Frame size	Voltage	Current	PF	Insu-lation	No of Rewdg

No load tests and Winding resistance

Voltage	Current	PF	KW	Ohms/phase as per RTC	Ohms/Phase as measured

Insulation Resistance-Phase to ground

R	Y	B

Standards

Insulation ResistanceIn megohms	<2	2–5	5–10	10–50	50–100	>100
Insulation level	Bad	Critical	Abnormal	Good	Very good	Excellent

Genaral Check up

Focus area	Remarks
Terminal board and studs	
Fan and its cover	
Shaft and Key way	
Shaft Key	

Rewinding Vendor Evaluation

Company name	Approval status	Comments on last rewinding	Comments on current rewinding

Tested by	Verified by	Approved by

6

Customer Focus

LEARNING GOALS

- **Types of Customers**
- **End-Use Performance Characteristics**
- **Customer Satisfaction Audits**
- **Coffee Stain**
- **Customer Satisfaction Index (CSI)**
- **Co-Creation**
- **Taylor-Made Products**

Introduction

"Customer is our boss" is an un-debatable quote in the case of any industry and it is more relevant in the case of spinning mills. To make an organization customer focused, we need to create a customer service environment and proceed with additional requisites to achieve customer satisfaction.

6.1 Types of Customers of Spinning Mills

Spinning mills customers may be listed as follows:

1. Knitters
2. Weavers
3. Value-added yarns like double/plied yarn manufacturers
4. Industrial thread manufacturers like cotton sewing threads/tyre cord-fabric manufacturers

DOI: 10.1201/9781003257677-6

Based on the type of customers, end-use requirements vary widely. Let us look into the end-use requirements of the aforementioned type of customers.

6.1.1 Knitters

Knitting customers are again divided into two major categories. One type of customer knits our yarn immediately into grey-knitted fabric and another type of customer directly processes the yarn before knitting. Processed yarn is taken for knitting. The first type of Customers are grey knitters and the second one is called as yarn-dyed knit fabricators.

End-Use Requirements of Grey-Knit Fabricators

- No shade variation within cone and between cones in a lot (should pass UV and natural light inspection)
- Yarn with a lower count CV
- Twist set yarn (by yarn conditioning or low-twist yarn)
- Higher fabric realization (99% and above)
- Less fluff liberation during knitting
- Absence of long-term faults like long thin and long thick
- Good running performance (no waste bunches/no objectionable faults)
- Absence of color fiber contamination
- No shade variation

End-Use Requirements of Yarn-Dyed Knit Fabricators

- Stronger yarn to meet the stress level during scouring/bleaching/dyeing
- Low level of hairiness as hairiness of yarn will be increased during dyeing/bleaching and subsequent winding
- Absence of color as well as white polypropylene/fiber contamination in yarn

6.1.2 Weavers

Weavers can be classified mainly into two broad categories as follows:

- Warping end use
- Weft end use

In the earlier days, the industry classified "weft end use" as low quality and whenever a yarn fails to meet "warp yarn requirements," the yarn will be sold to "weft end use." Today, with advanced technology, we have looms with a very high level of weft insertion rates, which require "a grade above warp yarn" in terms of running performance. Hence, this classification is of no use as of now. The next classification is similar to knitting end use, that is, grey fabric weavers and yarn-dyed weavers. The end-use performance characteristics for these two types of end uses are as follows:

End-Use Requirements for Grey Fabric Weavers

- Stronger yarn with less strength CV
- Cones with "nil" package defects
- Less contamination, especially color contaminations
- Less snarling tendency (especially in Airjet loom this leads to fabric defects)
- Better unwinding performance (market expects less than 0.1 breaks per million meters in warping)
- Less short thick and thin and long-term faults
- Exact or plus-length allowance in each cone to eliminate "run-outs" in fabric

End-Use Requirements for an Yarn-Dyed Fabric Weaver

- Very strong yarn with "nil" or nearer to "nil" weak places (as measured in Tensojet)
- Low level of hairiness and less CV of yarn hairiness
- Lower classimat faults especially short-thick faults

6.1.3 Value Addition Customers like Double Yarn/Plied Yarn/Gassed Yarn Manufacturers

These customers cannot change the impact of the parent yarn quality onto their value-added yarn and, hence, they are very selective while they choose the single yarn and their end-use process requirements are:

- "Nil" cone defects so as to have trouble-free unwinding
- No shade variation in "between cones in the lot" and "within cones"
- Less hairiness
- Stronger yarn with less strength CV

- Less strength CV
- Length per cone consistency as per actual cone weight and count

6.1.4 Industrial Thread Manufacturers Like Cotton Sewing Threads/ Tyre Cord Fabric Manufacturers

These product manufacturers expect the following Yarn quality:

- Length consistency in each cone
- Exact count and lowest possible count CV
- Stronger yarn with "nil" weak places
- Absence or less A2 and B2 classimat faults
- Splice appearance and splice strength should be very good

6.2 Customer Satisfaction

What is customer satisfaction? Can we infer the absence of complaints as customer satisfaction. Let us have a simple example to understand this phenomenon. We are going to a shop regularly to buy our groceries. In that particular shop, we experienced the quality of rice as poor, once. Suddenly, we thought about alternatives available for our next purchase if we don't have any relationship or loyalty to the grocery shop where we used to buy. But as a normal behavior, when alternatives are available we are tempted to try the alternatives and stick to the new vendor if the quality is better-whether it is at an equal cost or even at a slightly higher price. We *don't try to complain to the previous vendor that his quality is bad-which is the human tendency or buyers' behavior in 95% of the cases as per survey on behavioral pattern of customers.* Hence, absence of complaints doesn't mean that our customers are "satisfied with our product."

Then, how do we measure customer satisfaction? Customer satisfaction can be measured by:

- Repeat orders from the same customers
- Trend of Loyal customer addition annually
- Customer feedback-Only if the customer is interested to continue to procure from us, he will give feedback to us. Hence, customer feedback needs to be treated on top priority.
- Volume of sales improvement (not to account seasonal improvement in sales)

- Customer's new product requirements
- Open feedback on our product`s performance at their shop floor (end-use performance like Breaks in Knitting/warping, etc.)
- Periodical customer satisfaction audit
- Estimation of customer satisfaction index

(Model formats for customer satisfaction audits and estimation of customer satisfaction index are enclosed in Annexures XIX and XX).

6.3 Co-Creation

Co-Creation is a new concept of delighting the customers by working with them to manufacture the products which will satisfy their "end use performance characteristics and also meet the end products' quality requirements."

In spinning mills, there could be so many defined products as per their Spin plan and sales strategy. But when a customer asks for a particular type of product, we may need to understand their new requirements and translate into the changes in our Raw material requirements/Process parameter changes/Production levels/Cost of production etc. With an open mind, Spinning mills should discuss with the customers and work out the probable profit per kg also. After an initial discussion, we should create a sampling plan in discussion with the customers and samples must be produced and sent to them. Based on the samples, probable cost per kg can also be indicated. Customers' feedback must be received on the end-use performance and mills should make necessary changes in raw material/process parameters/Machinery parts(if there is a need) and start bulk production. Through Co-Creation, we achieve:

- More satisfied customer base
- More loyal customers
- Involved customer base with an understanding of our process and its capability
- Reduced manufacturing cost (No finished inventory/Cost of quality will be less with regular customers)
- Unique experience for the mills as well as for the customers

Co-creation becomes the order of the day for many leading spinning mills by offering more than 65% of their products through Co-creation.

Co-creation will sometimes be launched with the buyer audit of the spinning mills too. Most of the home textiles and knitted /woven garment manufacturers in India co-create their products through close relationships with their vendors-spinning mills and it leads to a very good "Customer Service Environment."

6.4 Taylor-Made Products

Taylor-made products are nothing but "products specifically made for meeting the end use requirements of the particular customer," which may or may not be acceptable to the other Customers of the same type. There will be certain customers who place orders with very specific quality specifications for their products, which will be totally different from other customers for the same product. An example of such a requirement is "Hosiery yarn with very low twist" for enhanced soft feel of the fabric. As the production of yarn with very low twist requires special care in Spinning (by reduced level of production) and Autoconers (by reduced tension during winding), which means a significant loss of production. Moreover such kind of specialized production also calls for a separate identification and channelization system along with clear work instructions to all the operators concerned to eliminate any kind of "mix up" of normal production with these special products. However, when a Spinning mill is ready to supply taylor-made products, there will be a list of specialized customers for the mills, which enhances the profitability and its consistency. Moreover, a loyal customer base is created through taylor-made products.

Nowadays, several mills are manufacturing taylor-made products with DNA traceability in imported varieties which is one of the best examples of taylor-made products.

CS 23: Case studies on: "Customer Service Environment-Take the initiative"

Spinning mills must train and create leaders who will take initiative in meeting the organization's objective of meeting the Customer's requirement. In one Spinning mills, discussion by the buyer and Top Management was in progress.

Buyer expressed interest in procuring a six-month requirement for 40s Combed Compact Hosiery slub yarn. Even though, the buyer is a loyal and longstanding Customer, sales team has refused to supply the same as it is not in their product profile.

At that time, the production head had suggested that this can be made possible by shifting the slub attachments from their carded yarn Ring frames where slub attachment is not fetching continuous orders. This was agreed by the Sales team as well as the top management immediately and buyer's requirement is fulfilled. This is made possible only by the ***initiative*** taken by the production head of the unit by his right-time suggestion.

Result: Customer has placed repeat orders as well as additional orders from this spinning mills and doubled his volume of procurement from this unit and became a regular buyer.

CS 24: Case studies on "Customer Service Environment-Customer Service Horror story"

Good experiences enjoyed by the customers help the organization to retain the customers for a longer time. At the same time, bad experiences by the customers keep the mills away from not only the concerned customer but also from other customers too.

One spinning mill used to supply 50s combed knitting yarn to the domestic market regularly. A new customer has procured 20 bags of the same from this mill and found the running performance is very poor with complaints of needle breaks in knitting fabrication. While trying to give feedback to the mills, the receptionist informed that she would ask the concerned person to revert immediately. At last, by dusk on that day, the Head of QA came on the line to the customer and enquired about the complaint. After hearing the same, Head of QA immediately replied that the product`s performance everywhere is good only and asked the customer to check on their knitting machine for the probable fault leading to Needle breaks.

Customer immediately sourced the information from his own Knitter's association and identified the knitters who recently procured the same type of yarn from these mills and the performance level from the fellow customers. Based on the performance details of this product which was found to be poor everywhere in the recent supplies even though the mill is in the list of preferred suppliers, he has taken up with the top management on behalf of all the customers, in person. As this is taken up at the high level, immediately the head of the manufacturing division with his team members investigated the whole process especially from spinning to autoconers/packing. The team identified that the lots under complaint were manufactured in the new Autoconer procured by the mills-before validation of the products by in house study. In the particular Autoconer Anti ribboning mechanisms were not set properly and "slough off" of layers occurred in every cone from this Autoconer, which caused needle breaks.

Had it been attended by the head of QA immediately, the issue would have been sorted out and the customer satisfaction level would have been improved. But unfortunately, the head of QA lost the opportunity to create a good customer service environment. In lieu of the hardships faced by the customers, this mill lost the customer base immediately. It took more than a year to regain the confidence level of the customers. Learning from this horror story is to "***Accept the voice of the customer and act immediately.***"

CS 25: Case studies on "Customer Service Environment-`Do more than the Competition`"

One Spinning mill used to supply 60s Combed Compact Weaving yarn to domestic weaving customers regularly. The weaving customer for this mill is a regular buyer for more than ten years and can be treated as a `loyal customer`. Once during his monthly sales meeting with the top management he mentioned that 60s Combed Compact warp yarn supplied to his warping division by another weaver(warping conversion) performed better than that is being regularly consumed by him. The performance parameters of both the yarns in their warping machine were reported as below:

TABLE 6.1

Warping Breakage Study

Mills	Warping machine efficiency	Warping breaks per million meter
Our mills	60%	0.6
Competitor mills	75%	0.35

Hence, the top Management deputed Head –QA to study the performance of both the products at warping shed to find out the root causes for the reduced level of efficiency and also the reasons for the breaks. Head QA has studied the warping performance of his mills and his competitor mills yarn in the same warping machine as a controlled study. By means of the study, Head QA identified that the warping breaks in his mills yarn is certainly more and analysis of break up of breakages revealed the following information:

- Initial layer breaks - 18%
- Stitches - 8%
- Middle layer disturbance - 40%
- Splice breaks - 12%
- Run outs - 12%

With this analysis, the Head QA had discussed with a cross functional team including Production, Maintenance and Electrical executives to sort out all these issues in the Autoconer. Once these issues were sorted out mills had sent fresh production to the warping shed and conducted the warping study once again. During this study, Head QA estimated the breaks per million meter is reduced to 0.2 (earlier level of 0.6) and the warping machine efficiency is increased to 82% (against earlier level of 60%. By doing all these corrective measures, mills have exceeded the level of performance of the competitor, thanks to the inputs provided by its loyal customer.

Annexure XIX: Model Format for Customer Satisfaction Audit

No	Queries	Weigh-tage	Details
1	Name of the unit		
2	Address of the unit		
3	Managing director		
4	Head of the unit and contact number		

(Continued)

No	Queries	Weigh-tage	Details
5	Products/Type of products-catering to which segment		
6	Monthly consumption in each variety of yarn		
7	Percentage of the volume of our yarn in each variety of their consumption		
8	Reasons for low or high percentage of volume in each variety of their yarn		
9	Immediate end use of our yarn		
10	Customer's rating on our yarn quality against their specifications on 1–5 scale(1-Poor/5-excellent) alongwith comments	20	1 2 3 4 5
10	Expected level of end use performance and their rating on our yarn	30	1 2 3 4 5
11	Competitors to our yarn and its performance level		
12	Gaps in performance level against competitors		
13	Customer rating on our delivery adherence	10	1 2 3 4 5
14	Packing mode of our yarn		
15	Customer satisfaction level on our packing in 1 to 5 scale alongwith their comments for improvement	5	1 2 3 4 5
16	Experience with our Sales team/Agents on communication on 1–5 scale alongwith their comments	10	1 2 3 4 5
17	Customer rating on our Customer complaints/Feedback redressal system	10	1 2 3 4 5
18	Current Procurement of yarn in kg/month 1. Vendor 1 2. Vendor 2 3. Our mills		
19	Sampling duration experience with our mills against others	5	1 2 3 4 5
20	Availability of Taylor-made products	5	1 2 3 4 5
21	Readiness level for co-creation	5	1 2 3 4 5
22	Feedback from Buyer Audits/External certification agency		
23	SPOC for future contacts(Single point of Contact)		
24	Anyother additional points suggested by customer for further improvements?		

Audited by:
Place :
Date :
Sign. :

Annexure XX: Estimation of Customer Satisfaction Index (CSI)

S.No	Factor	Rating	Weightage	Awarded marks	Remarks
1	Yarn quality	3	20	12	
2	End use performance characteristics	2	30	12	
3	Quality of packing	5	5	5	
4	Adherence to schedules	5	10	10	
5	Quality of communication by sales executives/Agents	5	10	10	
6	Effectiveness of our customer feedback/complaints redressal system	4	10	8	
7	Time taken for New samples	5	5	5	
8	Co-creation/Taylor-made products	5	5	5	
9	Co-creation of products	5	5	5	
	Total			72	
	Customer satisfaction index			0.72	
				Customer Satisfaction Index-Norms	
	Rating	1 to 5		CSI	Level of performance
	Excellent	5			
	Very Good	4		>0.95	Excellent
	Good	3			
	Average	2		0.85–0.95	Very good
	Poor	1			
				0.75–0.85	Good
				<0.75	Poor

7

Industry 4.0

LEARNING GOALS

- **Definition of Industry 4.0**
- **Evolution of Industry 4.0**
- **Goals of the Industry 4.0**
- **Technology in Industry 4.0**
- **Application Areas in Modern Spinning MIlls**
- **SMS-Smart Manufacturing Spinners**

Introduction

Industry 4.0 is a vision and journey. Organizations implement Industry 4.0 initiatives and prepare to turn the clearly documented Industry 4.0 vision, concepts, principles, technologies, and architecture into reality within their context. Germany launched the project "Industrie 4.0" concept to digitalize manufacturing in 2011 at Hannover. Moving beyond its roots, today there is a global transformation toward digital transformation of manufacturing and other industries.

7.1 Industry 4.0-What Does It Bring To Us?

The value created by Industry 4.0 far exceeds the single-digit cost savings that many manufacturers pursue today.

Industry 4.0 can be defined as the digital transformation of manufacturing through third platform technologies, such as Big data/Analytics and Innovation accelerators, Industrial Internet of Things (IIOT), Convergence of IT and OT(Operational Technology), Sensors and actuators, robotics, Artificial intelligence,

DOI: 10.1201/9781003257677-7

Smart decentralized manufacturing, Self optimized systems and Digital supply chain in the Cyber-physical environment.

This covers,

1. Simulation
2. Additive manufacturing
3. Autonomous robotics
4. Big data
5. Internet of Things/Industrial IoT(IoT/IIoT)
6. System integration
7. Cloud computing/Cognitive computing
8. Additive manufacturing
9. Cyber physical systems
10. Cyber security

We are in the early days of our maturity journey toward Industry 4.0 and every industry starts focusing toward the same. Projects around energy efficiency, factory energy management, and machinery maintenance management bring to us an entirely different world-of solutions/skill levels/standards. For example, additive manufacturing, robotics, or augmented reality are few projects every industry is working with. *Industry 4.0 has scope for the specialists who focus on "integration and Convergence."* Industry 4.0 requires a strategic view and staged approach.

7.2 Evolution of the Industry 4.0

First Industrial Revolution—Year of Start: 1784: First Industrial revolution brought mechanization, steam engines, power from water and steam, and new manufacturing systems in iron, textile, mining, metallurgy, and machine tools.

Second Industrial Revolution—Year of Start: 1870: Added technological electrification, mass production, globalization, Engines/turbines, and broad adoption of gas/telegraph and water supply.

Third Industrial Revolution—Year of Start: Was enriched with computer/internet/digital manufacturing/PLC/robotics/automation/digital networks/digital machines.

Now Industry 4.0-Year of Start: 1969: Calls for Convergence of IT/OT, autonomous machine, advanced robotics, Big data, Internet of Things

(IoT), digital ubiquity/smart factory machine learning, and Artificial Intelligence(AI). (Base: Cyber physical systems)

7.3 Goals of the Industry 4.0

Goals for the Industry 4.0 are automation, process improvement, and productivity/production optimization; the more mature goals are innovation and the transition to new business models and revenue sources with information and services as cornerstones. Industry 4.0 is sometimes called as "Smart Industry" too as it embraces "Smart manufacturing."

Vision of Industry 4.0 encompasses more than Automation and Data exchange in manufacturing technologies by

- *Stretching beyond technologies*
- *Looks into the end-to-end chain including warehousing, logistics, recycling, and energy conservation*

IoT-Internet of Things

Internet of Things consists of objects with embedded or attached technologies that enable them to sense data, collect them for a specific purpose (e.g., CCTV Data capturing inside the Spinning mills). Depending on the object and goal, this could be capturing data regarding movement, location, presence/absence of employees, work practices in shop floor, discipline of the shop floor employees, unsafe practices-the list is endless and voluminous. This data as such is a beginning, the real value can be appreciated only when this data is analyzed and acted upon.

IoT devices can also receive data and instructions, again depending on the end uses. In other words, what an IoT can do for the industry?

- *Track and trace possibilities: We can track the performance of a system/machine/men and get "accurate data online."*
- *Structural Health Monitoring: With the right kind of sensors and systems the structural health of all kinds of objects/production assets/cyber physical assets in Manufacturing and Industry.*
- *Examples: Ultrasonic detectors in critical zones to detect leakages and health monitors for critical machines like Ring frame, Combers, etc.*

IIoT-Industrial IoT

A key component of IIoT is connectivity. According to global research, industrial manufacturers still have to catch up in connectivity. Especially in India, many industries need to improve their connectivity. As per research in 2017, the number of IIoT connections across the globe would be 66 million. The current rate of new IIoT connections is 13 million per year, which is expected to be accelerated to *18 million per year by* 2021.

Manufacturers historically isolated their factories, plants, sites, and facilities from data connections. Today, significant opportunities are available to leverage the benefits of digital networks and enable extraction of data for analysis and ultimately improve "plant performance." In a nutshell, the benefits of IIoT are as follows:

- Improving the manufacturing efficiency
- Improving the machinery utilization
- Improving the productivity
- Enhancing employee safety in the industries
- Better service to customers
- Innovative process and product development

Additive Manufacturing

Additive manufacturing (AM) is an appropriate name to describe the technologies that build 3D objects by adding layer upon layer of material whether the material is plastic, metal, concrete, fibrous substance, or one day it could be human organs too...

Application of AM is endless. AM is being used to fabricate end-use products in aircraft, dental restorations, medical implants, automobiles, and even fashion products. AM extends itself to the diverse needs like:

- As a means to create customized product for consumers
- As industrial tooling
- Producing small lots for customers
- One day....production of human organs

IIoT-Technology of Connectivity

It moves at an accelerated manner from fixed-line connections to WiFi, 4G platform, and now toward low-power wide area (LPWA) technology. LPWA connectivity is gaining significant recognition as the dominant wireless connectivity solution for the IoT.

Cyber Physical Systems

Within the modular structured Smart Factories, cyber physical systems monitor physical processes, create a virtual copy of the physical world, and make decentralized decisions. Decentralized intelligence helps create intelligent object networking and independent process management, with the interaction of the real and virtual worlds representing a crucial new aspect of the manufacturing and production process. It helps Industry 4.0 to have a paradigm shift from "centralized" -to "decentralized" production. CPS are physical engineered systems whose operations are monitored, coordinated, controlled, and integrated by a computing and communicating core. A CPS integrates computing, communication, and storage capabilities with the monitoring/control of entities in the physical world from the nano world to large-scale wide-area systems—reliably, efficiently ,and in real time.

Application of CPS

- Healthcare industry: Medical devices on health management/ networks
- Transportation: Vehicular networks and smart highways
- Process controls in industry/spinning mills
- Production monitoring and controls in industry/spinning mills
- Maintenance management of assets (machinery and equipments) in industries/spinning mills
- Power consumption and energy management in industry/spinning mills

In general, "X" by wireless monitors and controls "Y," which are physical and wireless controls.

Trend: CPS has started its entry in a mega way by making 24/7 availability, 100% reliability, 100% connectivity, and instantaneous response even though it is complex. The benefits of the same are being fully utilized by every sector of the industry as of today and are being expanded manifold at a jet speed.

Application in Spinning Mills: One of the major breakthroughs in spinning mills is the development of Individual Spindle Monitoring systems(ISM) for Ring frames, which is based on this platform. This ISM has revolutionized the work culture at shop floor of the spinning mills, improved the analytical capability of the executives, pushed the managements for benchmarking themselves against competitors—dynamically—and also to take right kind of investment decisions. In a nutshell, CPS is going to be a vital tool in the hands of spinners who wish to become smart manufacturing spinners.

Cloud Computing

Cloud computing appears to have origins in network diagrams that represent the internet or various parts of it as schematic clouds. Cloud computing was coined for what happens when applications/services are moved into the internet "cloud." Cloud computing is not something that appeared suddenly overnight; in some form, it may be traced back to a time when computer systems remotely time-shared computer resources and applications. Many companies are delivering services from the cloud.

Example: Many Autoconer Management Information systems have been made available to spinning mills like Visual Managers from Muratec-Japan, which helps to scale against the best performers in the industry on key performance parameters in autoconers.

Characteristics of Cloud Computing

- Shared Infrastructure: Uses a virtualized software model, enabling the sharing of physical services, storage, and networking capabilities. The cloud infrastructure, regardless of deployment model, seeks to make the most of the available infrastructure across a number of users.
- Dynamic Provisioning: Allows for the provision of services based on current demand requirements. This is done automatically using software automation
- Network Access: Needs to be accessed across the internet from a broad range of devices such as PCs, laptops, and mobile devices, using standards-based APIs (e.g., ones based on HTTP). Deployments of services in the cloud include everything from using business applications to the latest application on the newest smart phones.

Benefits of Cloud Computing

- Cost Savings: Companies reduce capital expenditure.
- Maintenance: Cloud service providers do the system maintenance, and access is through APIs that do not require application installations onto PCs, thus further reducing maintenance requirements. SPMF-Smart Prescriptive Maintenance Framework—is developed on this platform.
- Mobile Accessibility: Mobile workers have increased productivity due to systems accessible in an infrastructure available from anywhere.

Big Data

Private companies and research institutions capture terabytes of data about their users' interactions, business, social media, and also sensors from processing machines/devices like their machines running in their factories, cars running on roads, cars on hire services, ambulance on emergency services etc. The challenge of this era is to make sense of this sea of data. This is where big data analytics comes into picture. Big data analytics largely involves collecting data from different sources, manage it in a way that it becomes available to be consumed by analysts, and finally deliver data products useful to the organization business.

Big Data Analysis Involves:

- Business process definition
- Research
- Human resources assessment
- Data acquisition
- Data mining
- Data storage
- Exploratory data analytics
- Modeling
- Implementation

Big Data-Core Deliverables:

- Machine learning implementation: This could be a classification algorithm, a regression model, or a segmentation model.
- Recommender system: The objective is to develop a system that recommends choices based on user behavior.
- Dashboard: A dashboard is a graphical mechanism to make this data accessible.
- Smart machines are already being manufactured, which are built on this platform by leading spinning machinery manufacturers. This will definitely help the spinners to move toward "Smart manufacturing."

7.4 Smart Manufacturing Spinners

Most modern mills in India are working with system suppliers to utilize almost all the systems described earlier and the examples of such systems are:

1. IoT-Internet of Things: Mills get data instantaneously from the running machines in a defined format in a structured format in mobile phones through WiFi. It helps the responsible decision makers to monitor and control the manufacturing operations remotely in time.
2. Big data analytics and Cyber physical systems: Examples are the systems which are making inroads in Spinning mills like:
   - Individual Spindle Monitoring systems: Smart spinners have already incorporated the ISM systems. With the help of ISM, these spinners are able to improve productivity (Kg per operative hour), production (Kg/hour), reduce Pneumafil waste as a percentage to spinning production, reduce fatigue of the operatives, improve spindle availability (reduction of idle spindles), identify rogue spindles, etc. Smart spinners are able to achieve minimum 3.5% production improvement in Ring frame with a range upto 8% in few mills.
   - Production monitoring systems: Production monitoring systems have been automated by drawing the data from sensors at vulnerable points to capture the accurate data from the machines. Data from the machines match exactly what is physically achieved by the mills. Hence, mills are able to plan their WIP/Despatch quantity in an accurate manner without any hidden surprises. Human intervention is eliminated for the estimation of actual production from the machines.
   - Energy management systems: Energy consumption measurement is essential to keep the energy cost under control. Energy sensors are used in the Smart series machines to capture energy consumption per unit of production continuously and we can take action based on real time data. For example, if two Ring frames running on identical parameters/conditions, there should not be any variation in energy consumption per unit of production. If it differs, then there is a scope to analyze and take action immediately. This helps the spinners to save energy cost through real time data, smartly.
   - E-bikes with dashboard: E-Bikes are used by the spinning operatives in Smart spinning mills where ISM is available.

Data from ISM is being captured in the dash board of e-bikes and operators need to act according to the status of each RF under his assignment. This reduces the fatigue of the operator, improves the mending time, and reduces the pneumafil waste in a Ring frame. Smart spinners have achieved a reduction of upto 60% with the help of ISM and e-bikes.

- Health monitoring systems: Machinery manufacturers also have utilized these concepts and have inbuilt newer technologies to monitor the health of the machines and alert the users in time. For example:
  A. Ultrasonic sensors for vibration montoring in Ring frame
  B. Thermal sensors in Ring frame main motor
  C. Ultrasonic sensors for air leakage detection in Autoconers

Conclusion

Excellence in spinning mills calls for Smart manufacturing by working toward benchmarking and scaling themselves up against newer limits through innovative practices in their organization. Definitely Smart spinners will be able to excel against the performance of their peers and sustain in their growth trajectory.

Annexure XXI: Smart Manufacturing Spinners (Industry 4.0)—Configuration

Smart manufacturing spinners need to incorporate the following equipments/accessories/systems in their new projects/modernization projects to excel smartly in manufacturing.

1. Well-connected smoke-detection and fire-detection sensors along with automatic quenching/diversion systems with alerts to responsible persons in the mills.
2. Health monitors to measure heat (temperature), vibration level in vulnerable machines/equipments, like:

- AWES-automatic waste evacuation sytems
- Humidification plant fan motors
- Comber headstock
- Ring-frame main motor/Headstock

3. Energy sensors in critical machines and equipments, like:
   - Ring frame
   - Autoconers
   - Humidification plants
   - AWES
   - Compressor
4. Air leakage measurement sensors in lap formers /Speed frames/Ring frames/Link coners
5. Humidity sensors in departments for automatic sensing department humidity levels and adjust the running mode of the supply fans/exhaust fans and pumps with suitable adjustments in dampers
6. Humidity sensors after eliminators to monitor the effectiveness of each plant
7. Dust-level sensors after filters in AWES/humidification plant exhaust system to prevent air pollution in the environment
8. Individual Spindle Monitoring systems in Ring frames
9. Spindle-identification(SP-ID) systems in Ring frames with Link coners
10. Visual management systems in Link coners to avail Smart Prescriptive Maintenance Framework (SPMF)
11. Latest Autoconer electronic yarn clearers with smart clearing facilities
12. Well-connected quality inspection instruments to give alerts on any deviation in key performance indicators

Mills can use IoT-enabled devices for getting alarms/alerts for any kind of deviation in performance levels.

We can use these devices :

- To stop the production in case of serious defect/deviation
- To divert defective products and continue the production after correction

- To alert concerned persons to attend the particular defect/spot
- To keep a record of events, which will be linked to knowledge bank and equipment history record

Conclusion

SMS—smart manufacturing spinners—can effectively implement the aforementioned recommended configuration toward "Smart manufacturing" and excel in the performance of the organization.

Annexure XXII: Payback Estimation—Factors To Be Considered for Installation of Individual Spindle Monitoring System

Individual spindle monitoring system is one of the first choice of any SMS—Smart Manufacturing Spinners, as it offers wide range of improvements in shop floor as listed herewith:

A. Rogue spindles reduction to the tune of 40%
B. Idle spindles reduction to the level of more than 85%
C. Reduction in "Slip spindles" (twist variation) by more than 50%
D. Improvements in Spindle-point production by a minimum level of 4%
E. Improvements in cop content by 8% to 10% as rogue spindles are eliminated
F. Pneumafil waste reduction to the level of 40% (reprocessing cost is eliminated)
G. Pneumafil motor energy savings with invertor controlled suction
H. Reduction in cop rejection level in autoconers

Based on the aforementioned improvements, mills can set new benchmark level of performance for :

A. Ring-frame utilization efficiency
B. Ring-frame targets for spindle point production in grams/spindle/8 hours
C. Increased work assignments for ring-frame tenters by a minimum level of 35%

D. Energy consumption in ring frame for every count
E. Pneumafil waste level
F. Hard waste level (as cop rejection is reduced)

Conclusion

With the benefits achieved in each of the factors listed earlier, mills can estimate their own payback for the installation of ISM—individual spindle monitoring systems—which currently varies from 6 months to 15 months. (This variation level is understandable as the base level of the performance of the spinning mills before the incorporation of ISM needs to be considered, which will not be uniform.)

List of Tables

(Continued)

List of Case Studies

Chapter No.	CS No.	Topic
1	1	Higher level of End breaks in 40s Combed Compact Warp yarn
	2	Higher Unevenness in Finisher Draw frame sliver
	3	Increase of Clearer cuts in Autoconers in 60s Combed Compact yarn
	4	Increase in Foreign fiber (FD) cuts in Autoconers
	5	Increase of Yarn Count CV in 40s Combed Hosiery Yarn during Final Lot Inspection
2	6	Impact of Buffing Accuracy on Yarn Quality
	7	Impact of Air leakage on Mis-splicing % in Autoconers
	8	Process FMEA on Weak places in 80s Combed Compact Warp yarn
	9	Impact of 5S implementation in a Spinning mill
	10	Design FMEA in a Spinning mills
	11	QAD-Maintenance interface
	12	MTTR-Mean Time To Repair (TPM)
3	13	Spindle point production improvement in 80s Combed Compact Weaving
	14	Impact of Short Stretch in Compact Spinning Ring frame
	15	Improvement of Spindle point production with Slim Tubes
	16	Utilisation Efficiency Improvement
	17	Yarn Realisation improvement
4	18	Improvement in work assignment in Ring frame through Re-training
	19	Improvement in work Assignment in Autoconer department through Retraining
5	20	Incorporation of MD Controllers
	21	Energy Savings in AWES - Automatic Waste Evacuation System
	22	Invertor control for Pneumafil in Ring frame
6	23	Customer Service Environment-Take the Initiative
	24	Customer Service Environment- Customer Service Horror Story
	25	Customer Service Environment-Do more than the Competition

List of Annexures

References

1. P. Chellamani, A. Kanthimathinathan, S., Karthikeyan (1991), Measures to Meet Yarn Quality Requirements for Export Yarn, Mill control Report no 7, Pages 10&11
2. D. Prakash Vasudevan, J. Jayaraman, R. Sreenivasan, Pasupathy (2019), Fiber Quality Index, SITRA Norms for Spinning Mills, Pages 7&8
3. T.V. Ratnam, K.P. Chellamani (2019), Neps and Fiber Damages, Quality Control in Spinning, Pages 7&67
4. Uster Insights (2013), Uster Classimat 5 Application literature
5. Indra Doraisamy, A. Chellamani, Kanthimathinathan (1995), Package Defects, Yarn Faults and Package Defects, Pages 44&57
6. T.V., Ratnam, P. Chellamani, A. Kanthimathinathan (1992), Maintenance Management in Spinning, Page 2
7. A. Kanthimathinathan (Mar/Ap 2021), Spin Solutions, Spinning Textiles
8. D.H. Stamatis (1997), Failure Mode and Effect Analysis, Risk Priority Number, Pages 25&183
9. J. Sreenivasan, T. Subash, Sambhaji, S. Chavai (2019), Costs, Operational Performance and Yarn Quality, Page 11
10. How To Assess Spinning Mills Productivity? (2019), Conversion Factors, Pages 17&35
11. T.V. Ratnam, Indra Doraisamy, S. Seshadri, R. Rajamanickam (1992), Cotton Cost Assessment, Cost Control and Costing in Spinning Mills, Page 1.5
12. G. Subramanian (1981), Part Analysis Training, SITRA Silver Jubilee Seminar on Training Textile mills Operatives, Pages 3.1&3.2
13. Elaine Biech (2011), Understand the Role of Trainers, 10 Steps to Successful Training, Page 18
14. Peter F. Drucker, Seven Habits of Effective Manager
15. A.R. Garde, T.A. Subramaniam (1987), Process Control in Spinning
16. Confederation of Indian Industry (CII) Publication on "Energy Efficiency Guidebook for Electrical Engineers"
17. Rahim Umer, Comparative Analysis of Uster Electronic Yarn Clearers of the Last Ten Years
18. Uster News Bulletin on UQ4 Clearers
19. Mercom Clean Energy and Insights, June 2020
20. GWEC (Global Wind and Energy Council)-Global Wind Report 2021
21. ET Eenergyworld.com on Renewable Energy
22. A. Kanthimathinathan, Scaling Us to Industry 4.0, Organisation Management Journal, Jan–Mar 2018
23. A. Kanthimathinathan Training and Development, Organisation Management Journal, Apr–June 2017

Index

Note: Page numbers in *Italic* refer to figures; and in **Bold** refer to tables

Printed in the United States
by Baker & Taylor Publisher Services